The American Technological Challenge

The American Technological Challenge

Stagnation and Decline in the 21st Century

Jan Vijg

Algora Publishing
New York

Library of Congress Cataloging-in-Publication Data —

Vijg, Jan.
The American Technological Challenge: Stagnation and Decline in the 21st Century / Jan Vijg.
p. cm.
Includes bibliographical references and index.
ISBN 978-0-87586-885-1 (soft cover: alk. paper) — ISBN 978-0-87586-886-8 (hard cover: alk. paper) — ISBN 978-0-87586-887-5 (ebook) 1. Technology and civilization. 2. Technology—History—21st century. I. Title.
CB478.V53 2011
306—dc23

2011023176

Front cover: Peter Paul Rubens, Vulcan forging Jupiter's Lightning Bolts. 1636–1638.

Printed in the United States

I dedicate this book to all the heroes of technology, from the anonymous inventors of agriculture, to Archimedes, Ts'ai Lun, Bi Sheng, Antonie van Leeuwenhoek, James Watt, Thomas Edison, Henry Ford, the Wright brothers and Douglas Engelbart.

Table of Contents

Prologue — Why I wrote this book

One of my earliest memories, probably from 1957 when I was three years old, is about the garbage truck slowly driving through our street. I remember how I wanted to be one of these men picking up trash containers and emptying them into the mysterious rear end of the truck. Every once in a while the inside of the truck seemed to turn around a couple of times, making all the trash disappear. Only later I learned that this was a rear-loader compactor truck — invented by Garwood Industries in 1938 — and that picking up garbage was not the ideal career my parents had in mind for me. I think about that often, during my commute in Manhattan when I am forced to wait behind such a rear-loader compactor truck blocking the street until similar men have picked up similar trash from similar streets to be compacted by hydraulic cylinders similar to the ones that at the time revolutionized garbage collection.

Patiently waiting, it naturally occurred to me to wonder why in the largest city of the most powerful nation on earth seven decades were not enough to develop a more advanced system. After all, when the Romans could build their great central sewer, the Cloaca Maxima, more than 2,500 years ago, you would expect that we could build an automatic trash disposal system based on conduits, a central vacuum system and remotely located trash receptacles. After all, technology is continuously improving — or is it?

When I grew up during the sixties the year 2000 represented a magical new world and a symbol of everlasting technological progress. The feeling associated with the year 2000 and beyond is symbolized by the television series 'The Jetsons'. With their homes and workplaces raised high above the ground on poles, their transportation limited to private flying vehicles, and surrounded by a host of machines to make life easy and above all leisurely, the Jetsons represented what we all expected of the future. This was a time when confidence in science and technology reached dizzying heights. Progress was taken for granted and the general mood that technology knew no boundaries is epitomized by science fiction, which has never been more popular than in those days.

Alas, writing from the first decade of the 21st century such a golden age seems more distant than it was then and the conclusion that we will never be able to emulate the Jetsons is all but inescapable. This is perhaps best illustrated by the disappointing progress in transportation. More than anything else, novel forms of transportation have been the face of 19th and 20th century progress, from the steam locomotive to the airplane and the first men on the moon, immortalized by much of Jules Verne's work and symbolized by Filippo Tommaso Marinetti's racing motorcar in his Futurist Manifesto of 1909. I could not help thinking about that when early in 2009 Tata Motors of India unveiled the world's cheapest car — the Nano. Of course, more than half a century ago Fiat introduced a very similar car, the Cinquecento, which gradually replaced the scooter for millions of Italians who found themselves in the midst of their own economic miracle after the devastation of the Second World War[1]. By now you would think that our transportation sector might come up with a more creative solution than a gas-driven, manually controlled vehicle.

1 The Cinquecento, Italian for 500, was phased out in the 1970s. A modernized version of the same car was introduced in Europe in 2007 and in the US in 2011.

Somewhere in the second half of the 20th century, progress in transportation stalled. Apart from the three decades when the supersonic Concorde regularly broke the sound barrier, the speed of travel has not substantially increased since the de Havilland Comet opened the first commercial jet service, from London to Johannesburg, in 1952. This disappointing development was unpredictable in the 1950s and 60s when it was generally assumed that supersonic passenger transport would soon replace the more mundane jet planes, which themselves had previously replaced propeller planes. And, yes, in the 1970s the Concorde became the first supersonic passenger airliner to enter service. A product of British-French collaboration, the Concorde is now regarded as an aviation icon. It has the dubious honor to be the first plane in aviation history to end its tenure without a superior successor!

The 1968 movie *2001: A Space Odyssey*, directed by Stanley Kubrick, was rather conservative in anticipating progress in human space exploration. Instead of traveling through hyperspace or some other predicted physical breakthrough to accomplish faster than light travel, the movie contains some lovely scenes of a Pan Am space shuttle bound for a permanent international space station and, eventually, a lunar base, all visualized with painstaking realism. Now, 10 years after 2001, the first international space station (significantly less impressive than in the movie!) is barely completed with the space shuttle in the process of being retired. Indeed, rockets very similar to the original Vostok 1 that brought Yuri Gagarin eternal glory as the first human in space in 1961 are still our only option to escape the gravity of the earth. Six moon landings from 1969 to 1972 had no successors and have not given us a lunar base.

Most of us take technological progress for granted. And, indeed, better cars, improved cell phones, smaller and more energy-efficient appliances, more powerful computer chips, new drugs and smarter ways to implant medical devices, and lots and lots of books, magazines and movies with the latest in mind-boggling inventions, surely strengthen us in the belief that our scientists and engineers are infallible in turning out ever better tools and processes to make our lives increasingly comfortable, safe, pleasant and exciting. As a biomedical scientist trying to make sense out of the living world in the hope that this will someday contribute to curing disease and extending healthy life span, I have always been interested in the question where science and technology would bring us. As a teenager I was convinced that I would soon be witness of extensive human planetary explorations, the transformation of our cities into gleaming, compact hubs to be reached in hours from every point on the planet, cures for cancer and other diseases, and much, much more. And I must say, my life as a scientist has been extraordinary. I have had and still have the privilege of witnessing some of the most remarkable events in human history: deciphering and understanding the code of life as this is embedded in the molecules we now all know as DNA.

Naturally inclined to be optimistic, I cannot fully remember the moment when I first realized that something was amiss. It might have been a few years ago when I realized that the airplane that brought me from Amsterdam to New York was not very different from the one I took in 1984 for my first visit to the United States. The car that I now drive is very similar to my first car, the one that I bought in 1988. The subway trains and stations in New York do not look much different from when I first saw them in 1984. Even the personal computer, long our pride and glory, has not changed much since I bought my first one in 1986. Even in my own field of work, the biomedical sciences, all the amazing progress in our understanding of the living world no longer seems to give us breakthroughs in clinical practice, with the latest drugs for combating human disease merely slightly improved versions of earlier ones. Were it not for the internet and the worldwide web I would say that our lives have been pretty stagnant during the last decades.

It took me a while to realize why I was actually uncomfortable. While I was clearly disappointed that the dreams of my youth had never materialized, neither me nor anyone else I knew seemed to have given up on them. Like everybody else I was convinced that technological progress was faster than ever and that we would go to Mars, harness nuclear fusion, build smart highways and soon reap the fruits of household robots, inexhaustible batteries and faster airplanes. But why did these things always remain just around the corner? Had it always been like that? Suppose I would have been born 50 years earlier, i.e., in 1904. That would have allowed me to witness the rise of the motor car, the birth of modern air transport, almost all modern household appliances, the agricultural revolution, nuclear energy, the transistor and the computer, antibiotics, television, the electron microscope, the photocopier, the kidney dialysis machine, electronic hearing aids and, yes, the compacting garbage truck and much, much more. My world would have transformed before my very eyes!

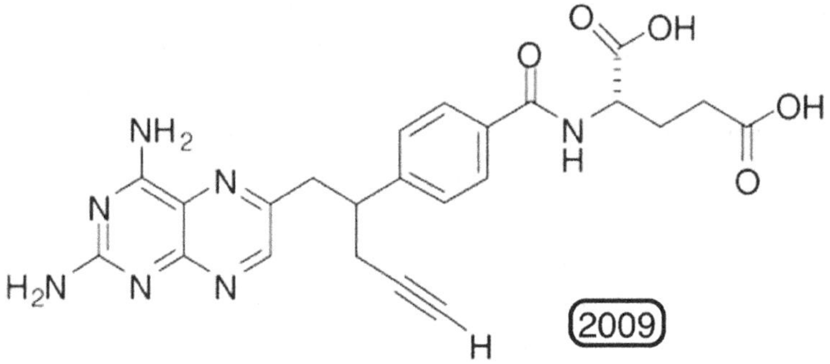

2009

Pralatrexate (PDX)

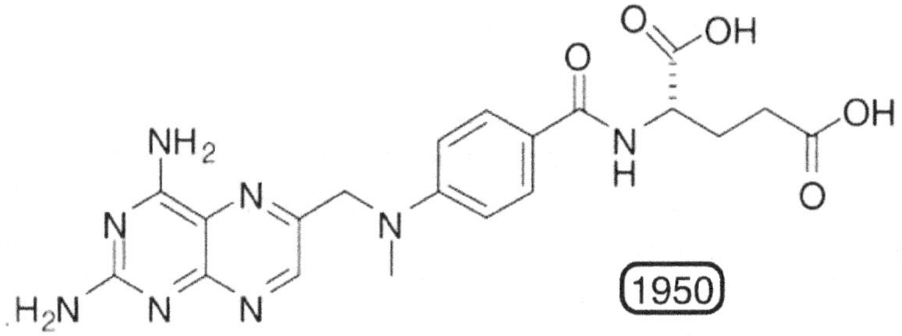

1950

Methotrexate (MTX)

Born in the 1950s I am left with my own telephone in my pocket and the internet, not enough to be awestruck with wonder. Yes, it is true that myriads of improvements have greatly increased efficiency and reliability of all these earlier inventions, but that is not the same as the wonderment aroused by something truly new and life-changing. Considering the possibility that technological progress could actually be slowing down rather than progressing ever faster I began to discuss this with friends and colleagues. Most of them thought I was wrong and without exception pointed towards the tremendous progress in information technology as dictated by Moore's law — the doubling approximately every two years of the number of transistors that can be placed on an integrated circuit. The argument did not convince me. Indeed, while Moore's law is often used by futurologists to describe the exponential nature of technological progress in general, it really only tells us about our capacity to significantly improve an existing invention. It does not say anything about progress in information technology, such as the creation of intelligent machines that can understand and communicate with us. Instead, we still have to deal with Microsoft Windows and the like. Not much progress since the 1980s! Rather than placate my concerns, these discussions merely intensified my feeling that for some reason we were no longer making the great strides in developing and adopting new technology we have become accustomed to.

As a scientist I am naturally interested in technology, which has always been critical in driving progress in science, from Antonie van Leeuwenhoek's microscopes to the Large Hadron Collider, the world's largest and highest-energy particle accelerator. When I gradually began to study the problem in more detail many questions came up. What exactly is technology and how does it relate to humans? What are the incentives for humans to develop new technology and how do they do it? Why are some societies innovative and some not? What determines changes in innovativeness in society and can this be manipulated? Because nothing is as good a guide to analysis as converting thoughts and reasoning to ink on paper (or bits in silicon), I began to write this book.

I quickly realized that inventions themselves are only one of the factors that determine the technological fate of a society. Sometimes, inventions are adopted, and eagerly; sometimes not. Therefore, I realized that inventions must be analyzed in the context of society, and societies cannot be studied in one dimension. They are the products of history, another of my favorite subjects. The history of technological progress, and the historical and societal factors that impel or restrain the adoption of inventions, are explored in the book.

The book is divided into two parts. In Part One we will take a survey of technology in our current, global society that has its origin in the industrial revolution in Europe at the end of the 18th century. Chapter 1 introduces us to technology as the critical factor in our economic well-being and the main driver behind humanity's paths to prosperity. The downward trend in innovativeness, suggested by the numerous anecdotal examples, is confirmed by the scientific literature as well as by simply counting the numbers of major inventions listed in Wikipedia, the encyclopedia freely available on the internet. Also my own research of identifying over 300 well-defined, major inventions since 10,000 BC (10,000 years before the beginning of our common era, i.e., the birth of Christ) indicates that far from growing exponentially our inventiveness, i.e., the number of new inventions per unit of time, has reached a plateau sometime in the 1970s and may actually find itself in a downward spiral. This is unexpected and was certainly not anticipated by our immediate ancestors from the 20th century, who became the prime beneficiaries of the fruits of the industrial revolution. Most of the predictions made by futurologists as late as the 1960s did not come true. Why this happened is the main question addressed in this book.

In Chapters 2–5, we will take a closer look at stalling innovation by evaluating progress in four major areas of technology development, energy, transportation, medical technology and information technology, against the backdrop of the bullish expectations of half a century ago. What did previous generations expect us to accomplish? How did they see their future society? Were those views realistic and why are there only so few cases that would have met their expectations? Why was the steamboat so readily accepted in the 19th century and the Concorde a failure in our own day? Why could we develop a moon-landing program within a decade and regular space travel remains a utopian dream? Why did we so brilliantly beat back the scourge of infectious disease early in the 20th century only to lose the war on cancer? Why were the 19th century visionaries, such as Jules Verne, largely proved right, while the stargazers of the 20th century (Isaac Asimov, Arthur C. Clarke) had it wrong?

It appears that there certainly are ideas as great as in the past, but that current inventors experience increasing difficulty in seeing their inventions adopted by society. The various technologies that we have come to rely on are strongly integrated in common practice. To see them replaced by new ones, especially potentially transformative inventions, proves much more difficult than the implementation of the original technology in the first place. For example, it proves to be much more difficult to implement smart grid technologies than getting the original electrical grids accepted now more than a century ago. And we have already seen that the first supersonic passenger aircraft, the Concorde, never made it, mainly because the airlines had no interest in adapting their established flight schedules and simultaneously fight off complaints about high noise levels.

Naturally, many readers will now wonder why, if technology development is really slowing down, it does not seem to do that in the electronics industry. As they will argue, new electronic gadgets and internet applications appear almost on a daily basis. Indeed, the information revolution, in the midst of which we currently find ourselves, is conclusive proof of accelerating technological progress. Or is it? First, it should be clear that much of the progress in this area of technology development is real and also fairly recent. However, alone this cannot compensate for the slowdown in other areas and much of the progress in information technology is focused on entertainment, social networking and financial tools. While such progress is certainly innovative, it reflects a growing imbalance between what we can call hard and soft technology. Our society is characterized by a shift away from science and engineering towards finance and entertainment. This is driven, at least in part, by frustration experienced by the increasing resistance of society to the implementation of inventions in areas such as medicine, agriculture, energy and transportation. Developing new tools for digital entertainment, social networking or short selling stock typically does not lead to angry protesters or excessive government regulation!

Chapter 6 systematically discusses the possible causes of our current technology slowdown. For example, one could argue that there may be a natural limit to human progress. Or, that there are not that many brilliant new inventions anymore to be made and more technological progress simply requires much greater effort. These arguments, which have been made many times in the past but were always overruled by the reality, sound reasonable and may even to some extent contribute to our current technology slowdown. However, as pointed out in the previous chapters, there are lots of new, breakthrough inventions imaginable and the evidence simply does not suggest that further, dramatic technological progress is not within our means. There are no hard facts suggesting that we have reached the end of the possible. Indeed, there are lots of examples of great inventions waiting for society to give them a chance. However, we may have become too successful to give them that chance! Rather than the chaotic and violent society of our immediate ancestors, always ready to adopt the new without giving a second thought to possible failure or dangerous side effects, we are full of complicated regulation, public resistance, perverse profit motives, vested interests and standardized business models. This effectively prevents us from adopting new, groundbreaking technology.

Difficulties in implementing new technology explain the current shift towards short-term, directed research in industry and an incremental improvement approach to technology. I would call this the decline of American dynamism, the focus of its industries on short-term profits rather than the farsighted projects of the past, out of fear that mutual fund managers will dump their shares. Not coincidentally, this increasing 'hedge your bet' mentality occurs at a time when world society is changing for the better. Everywhere people live longer, getting richer and are less likely to become a victim of war and violence. Naturally, this is a consequence of increased regulation, more accountable governance and the ascendance of rational thinking in international relations. Paradoxically, it is this same rational stability that now holds us back from making the same dramatic leaps of progress as in our less rational past. In short, what I believe is the true cause of our technology slowdown is the very success of our society in creating a brave new world where everyone is safe and satisfied enough not to be tempted taking on too much risk.

This begs the question: Is ours the first society suffering technology decline? Could it be that technology ups and downs are recurrent events in human history and not an atypical phenomenon limited to our own age? There is a tendency to see world history as a monotonous increase in human success in manipulating their environment. In reality, however, planet earth most of the time was occupied by different societies, overlapping in time and often unaware of each other's existence. Indeed, there have been highly successful civilizations that went before us and were also based on superior technology, from the first hunting and fishing tools to agriculture, writing and monumental architecture. This is the topic of Part Two, in which we will see that our current technology slowdown appears to be just an example of history repeating itself.

First, in Chapter 7, I will trace human evolution and early history to see how technology and innovation are inextricable parts of our nature, and how waves of technology development have been responsible for the many successful societies on our planet since the origin of *Homo sapiens* almost 200,000 years ago in Africa. Most histories of technology stress overall human progress, culminating in the industrial revolution and our current global economy. By contrast, I will focus on the diversity of humankind and sketch the contours of not one but multiple, parallel histories of technology. It appears that historical patterns of technology are characterized by recurring episodes of advances followed by stasis. I will identify the common factors associated with periods of technological brilliance and the factors that cause decline.

Applying the lessons of the previous chapter, Chapters 8 and 9 zoom in on the two most successful societies that went ahead of us, i.e., the Chinese and Roman Empires, and the remarkable similarities with our own society in the rise and decline of technological progress. Both ancient empires were born from a period of turbulence and internecine warfare between highly competitive states. It was the unification of these states by one hegemon that brought peace and stability. Yet, it was the period of turbulence, not the period of stability that proved the most creative in generating technological progress. The similarities to our own 21st century society are striking. At the height of their power, when their states had stabilized, with their citizenry as safe and protected as never before, with extensive bureaucratic governance attempting to regulate every aspect of society, innovation in both the Roman and Chinese Empires had irreversibly stalled. Like with us, their very success seemed to have bred stasis, preventing further technological progress.

But it would be a mistake to think that the lack of technological progress was also the recipe for decline. Technological progress had stalled well before these empires had reached their apogee and there is no evidence that either the Roman or Chinese Empire collapsed for any other reason than foreign aggression. Had they made further technological progress and, for example, invented the machine gun, they may still exist. But the fact is that they did not, because their own stable and successful societies no longer provided an incentive for quickly adopting new, breakthrough inventions. Can we now see our future through their past?

Chapter 10 provides the background of Europe's rise to dominance. I try to explain how after the collapse of the Roman Empire in the West a new society of peasants, knights and priests managed to take advantage of their unique geographical position. We will see that Europe's success, like that of Rome and China was primarily based on fierce inter-state competition. Chapter 11 shows that this also eventually resulted in Empire, first the

British Empire and then its successor American Empire. Like China's and Rome's early empires, these empires were initially also not stable. Stability only came at the end of the great wars, the First and Second World War and the Cold War. Now, in 2011 we can see, for the first time in history, how a world empire of nation states is growing out of the world's first true global economy. Similar to what happened in the Chinese and Roman Empires also the success of this stable world oikoumene is associated with declining innovation and the dampening of technological progress.

Finally, in the Epilogue I will attempt to predict the consequences of our current technology slowdown. If at different times in different societies the fundamental causes of technology slowdowns are similar, will our own innovation crisis be fatal to us and lead to our demise? Are we on the long road down, on our way to a de-industrialized future? Will we decline like the empires that preceded us? How will a lack of major technological advances affect our wealth? Can we still thrive over the next few centuries, even without major new inventions? Most importantly, are there ways to rescue us from technological oblivion? The answer that I offer is that in contrast to great civilizations in the past, ours is a truly global one without external threats. Successful societies before us did well for centuries in the absence of further technological progress until they succumbed to foreign encroachment. Since ours is a true planet-wide economic empire, we may well remain successful, albeit at some cost. Indeed, the pain from no longer making great progress in technology is felt precisely in our wallets. In the US, for example, the personal income of the average citizen has remained relatively stagnant over the past few decades. There are many barriers to be overcome before we can hope to see dramatic increases in wealth generation again. Accomplishing that feat, if at all possible, requires massive, public–private partnerships.

Finally, now that I have made it clear why I wrote this book, I also want to clarify what it is not concerned with. While I am aware that in what follows I occasionally show my own personal appreciation for the new and the revolutionary, this book does not have the wider aim of convincing the reader that continuous, dramatic technological progress is necessarily always a good thing. My purpose is not to give judgments on the many arguments of why certain technologies should not be adopted. I am not disputing the validity of safety issues, ethical or economic constraints or even irrational resistance. The fundamental aim of the book is to help myself to understand why technological progress has lost so much of its luster over the last decades. Technology is an integral part of human nature and defines us to a large extent. The main conclusions of this book are that insight in the interaction of technology with society can help us mitigating or resolving problems and shape our future. A closer understanding of the ties underlying this symbiosis will help us to know what lies ahead.

Part One — Broken Promises

Chapter 1. Retro Technology: The Facts of Innovation

Some time ago, when walking through a typical big-airport mall, my eyes were caught by a sign for 'Technology'. The store advertised in this way offered a wide range of the latest and most exclusive phones, laptop computers and a dizzying display of electronic gadgets. I realized that in the minds of many, technology is the sleek iPad, symbolizing the tremendous progress in information technology as the latest wave in the series of breakthroughs that transformed society since the beginning of the Industrial Revolution now more than two centuries ago. The evidence is all around us, from Amazon, eBay, Google and Wikipedia to YouTube, Facebook and Twitter. For us, consumer electronics are what steam engines and trains were for our 19th century ancestors. Can there be any doubt that we live in an era of increasingly rapid innovation?

Progress in technology is taken for granted, but few realize its critical importance to our well-being. Technology can be defined as the tools we use to make our lives more comfortable, drive our economy, provide our food, clothing and shelter, help us to communicate, cure our diseases and give us leisure. We call ourselves *Homo sapiens* or thinking man. But the armchair in which we do the thinking is technology. The papyrus on which Plato wrote his *Republic* is technology. And the computer I use to type this text is technology. Technology is us!

One could imagine how the first generation of missile weapons dramatically changed the playing field where our primitive ancestors had to hold their own against an often-hostile environment. Inventions as the javelin, sling, and bow and arrow suddenly allowed humans to begin dominating the animal world, finding protection against violent attack and secure a food supply through hunting. In turn, this created leisure time, time to enjoy life and paying attention to things unrelated to continuous foraging and watching out for predators. This was the time when, similar to our current entertainment industry that grew up with the invention of the personal computer and the internet, early humans developed leisure factories of their own. The impressive works of art created by the Cro-Magnons in the caves of Southern France 30,000 years ago are silent reminders of that.

The power of technological progress has never been on display more convincingly than in food production, obviously an important aspect of human life. The invention of agriculture, now more than 10,000 years ago, for the first time permitted food production and storage on a large scale. This transformed human society from sparsely distributed, cave dwelling nomads to much more densely populated village cultures and ultimately the large populations of big ancient cities as Ur, Alexandria and Rome. However, dearth and famine remained facts of life. Indeed, at the end of the 18th century the British economist Thomas Malthus predicted that every gain in food production would lead to a population increase, which in turn would inevitably give rise to food shortages in bad times and a subsequent population decline.[2] He formulated this as an iron law, based on a linear growth in food production (1, 2, 3, etc.), but an exponential increase in population (2, 4, 8 etc.). The predicted crisis never occurred because technological changes in agriculture (machinery, fertilizer, pesticides) managed to secure the human food supply irrespective of the magnitude of population growth. And our cities grew into mega-cities with some now more than 20 times the size of ancient Rome (or the London of Malthus, for that matter).

Malthusian crises can affect food or other commodities and they have been predicted as recent as 2008. In July of that year my wife and I drove from San Francisco to New York and quickly found out that one tank of gas cost us almost $100. European prices, indeed! The reason resided primarily in record high prices of oil, which were almost $150 a barrel in that month (compared to less than $50 four years earlier). These rising prices were due to an increase in global demand for petroleum, which in turn was a conse-

2 Malthus T.R. 1798. *An essay on the principle of population.* Chapter 1, p. 13 in Oxford World's Classics reprint.

quence of the successful completion of the first truly global economy, with a huge increase in demand from former laggard countries, like India and China. Again, like in the 1970s, when political turmoil caused the so-called oil crises of 1973 and 1979 with suspended exports and reduced production, predictions of a definite end to fossil fuel use made the rounds. This time, however, the cause was not political but a matter of limited supplies, unable to keep up with the rising demand. But as we had seen before, such shortages quickly came to an end. New technology in the form of a revolutionary way of drilling for oil and natural gas from vast formations of shale rock began to unlock potentially gigantic new supplies, especially of natural gas[3], but also oil[4]. This went so quickly that while in the US construction of new terminals for receiving liquefied gas to be imported from the Middle East and elsewhere was already underway, 'horizontal fracking' as it is called turned the US from an importer to an exporter of natural gas almost overnight. Prices promptly went down to the benefit of the consumers[5].

Dire predictions of looming catastrophes are nothing new. We saw them in the European Middle Ages when the year 1,000 was considered to be the year when corrupt society would be destroyed and replaced by the Kingdom of God on Earth, lasting a thousand years or more. And we have seen it not that long ago, when in 1973 a group of wise men, the Club of Rome, published their report, *The Limits to Growth*, warning against unlimited and unrestrained growth in material consumption in a world of clearly finite resources[6]. In a similar vein, Jared Diamond, in his book *Collapse: How Societies Choose to Fail or Succeed* warns for threats to a successful society's survival, mainly under the influence of environmental problems, such as climate change[7].

Being an incorrigible optimist, I have the tendency to take such doom scenarios with a grain of salt. Not because they are factually wrong: in their

3 Schulz M. "The Quiet Energy Revolution", *The American*, February 4, 2010

4 Konigsberg E. "Kuwait on the prairie." *The New Yorker*, April 25, 2011.

5 Natural gas prices are not yet determined by the world market, which explains why oil prices are not affected by an increase in domestic production. In the future, however, because of an increase in worldwide trade of liquefied natural gas North American natural gas markets are likely to become increasingly dependent on the world market price as well.

6 Meadows DH, Meadows DL, Randers J, Behrens III WW. *The Limits to Growth*. Universe Books, New York, 1972.

7 Diamond J. *Collapse: How Societies Choose to Fail or Succeed*. Viking Penguin, New York, 2005.

time and culture the millenarians made a lot of sense, because the early European Middle Ages were far from a pleasant time to live; the Club of Rome was correct in virtually all the points they made; and Jared Diamond based himself on concrete examples when warning for societal hubris that has ended many a successful society. I am optimistic because millenarianism subsided when Europe went through its first major technological revolution around the turn of the millennium, technological progress effectively prevented the scenario sketched by the Club of Rome, and Diamond's warnings of impending disaster, when heeded, are undoubtedly addressable by yet another set of technological changes. There is abundant evidence that our society is the most successful one ever and I am a strong believer in the power of the human brain to get us out of every possible mess, including the ones we ourselves have created.

Technology is the wellspring of advances in the human condition, from the invention of the first stone hand axe to our current repertoire of computational and telecommunication tools. Nobody would deny that further progress in technology development is essential to guarantee continuous increase in wealth and quality of life, which we all seem to take for granted. Most of us feel that such progress is indeed forthcoming, and we generally do not doubt the capability of our scientists and engineers to provide new and improved tools at an ever accelerating rate. With that mindset, the thesis of this book, i.e., that historically high rates of invention and innovation are on the decline, is not only a disconcerting but a ridiculous proposition, one to be rejected out of hand. In fact, futurologists have come to see technological progress as ever accelerating, based on the argument that each invention spawns many other inventions.

And it is true that a lack of technological progress is probably the last thing that goes through your mind when downloading the latest apps for your iPhone, ordering an eBook Reader or follow instructions from your GPS navigation system to your favorite restaurant. However, without trying to diminish progress in information technology, much of which is real and does reflect major new inventions, accomplishments in this area overshadow disappointing progress in other areas. This is not so easy to see because, in spite of the many overhyped cases that went nowhere, real progress is still being made. We all know that improved microprocessor chips continue to appear, electrical appliances have never been more energy efficient, new broadband services become cheaper, faster and more versatile every month, the latest heart valves can now be implanted through catheters rather than by traditional open-heart surgery, and with the first electric cars on the market there is hope for a more environmental-friendly world.

But on the other hand, it should be obvious to those who approach retirement age that the world is no longer changing as fast as it used to do. Today, the word processor I use to write this book is hardly better than the one I used 20 years ago, we drive the same cars, fly the same jetliners and live in the same cities as in the 1960s; we are still unable to cure cancer or Alzheimer's disease, and our rockets are still not sophisticated enough to enable excursions to the moon and planets. We are still using coal to generate electricity, gasoline to fuel our cars, the sewing machine of 1830 for sewing clothes and have come to accept hours of battery life rather than the nifty inexhaustible power packs promised to us in what now seems a long time ago. The long lines at airport immigration have not been dissolved by the general introduction of instant biometric identification and we have never seen the smart cars and highways that should by now have resolved the traffic jams.

These are of course merely examples. However, most of us would probably agree that the rate at which major inventions penetrate society has become disappointingly slow as compared to the previous century. But while there may not be as many breakthrough inventions, we do make lots of improvements in existing technology. How does that differ from the past? Are we really witnessing a decline in technological progress or do we merely look through a distorted prism? Therefore, before we seriously begin to analyze different areas of technology development, trying to understand why progress suddenly is no longer as fast and dramatic as it often was during the last two centuries it is important to get away from anecdotal evidence and look for objective indicators to see if my sense of discomfort with current technological progress is warranted.

Measuring technological progress

Technological progress is based on inventions, i.e., new creations such as a tool, a device or a process resulting from ideas and usually some experimentation. Note that an invention is not the same as scientific discovery. While scientific discovery, the gain of new knowledge about us and the world around us, is not associated with utility, inventions always are.

Sometimes (but not always!) scientific discovery can drive inventions and therefore the creation of new technology, but they are not the same thing.

The act of invention is not sufficient to earn eternal fame as many an inventor has experienced. It often takes a long time for an invention to gain the recognition that it is potentially useful and sometimes even longer before it is applied on a large scale. The process of making inventions useful to society is called innovation. It is innovation that translates inventions into increased economic growth and social well-being. Inventions and innovations are often used synonymously and occasionally I may do that too. I will also use the term 'innovativeness' interchangeably with 'inventiveness' to indicate a society with a high rate of developing and adopting new, transformative inventions. As will become clear in this book, the willingness of a society to adopt new inventions is as important as a climate that fosters making such inventions in the first place. In fact, when an invention remains un-adopted chances are it will never even be known as an invention. The process that leads from a potential invention through innovation and adoption to the realization that technological progress has been made is schematically depicted in Figure 1.

Lately, several columns in newspapers and magazines expressed concern that the US is about to lose its pre-eminent position as the most innovative state on earth. It was argued that we are suffering from an innovation shortfall and too few breakthrough inventions, with the economic boom of the 1990s merely built with borrowed money. A decline in American technological progress is also the thesis of a 2010 book by Lynn Gref, *The Rise and Fall of American Technology*. All of these authors have noted that ever more technology products are no longer made in the US but in Asia[8].

It is certainly true that Europe and the US will lose their preeminent position as the economic centers of gravity to Asia. But this is simply a consequence of the inevitable process of that part of the world catching up. Asia's success is built upon huge population centers of highly civilized, traditionally hard-working societies. Why wouldn't they reach economic parity with Europe and the US and, based on their sheer size, eventually dominate the world economy? Nevertheless, I wonder if Asia is doing any

8 Miller CC. "Another voice warns of an innovation slowdown". *The New York Times*, September 1, 2008; Mandel M. "The failed promise of innovation in the US." *BusinessWeek*, June 3, 2009; Gref LG. *The Rise and Fall of American Technology*. Algora Publishing, New York, 2010; Phelps ES. "The economy needs a bit of ingenuity". *The New York Times*, August 6, 2010.

better, technology-wise. After all, they make what the US and Europe used to make but cheaper and often better. While such colossal newcomers as China and India may have the world's most dynamic economies and soon lead the pack of super states, I have not seen major inventions coming from that direction recently.

Indeed, innovation problems are very much on the mind of Asian leaders as well. In China, with its Communist Party-controlled, top-down policies, the need to come up with the world's next breakthrough products is keenly realized. Following the reasoning that those nations that file the largest number of patents generally also have the most innovative corporations and Nobel prize winners, China wants to double the number of patents that its residents and companies file in their own and other countries. To make that possible and increase the number of patents from about half a million in 2009 to 2 million by 2015, they provide incentives, such as cash bonuses and tax breaks[9]. The question of course is whether by such methods a state can engineer itself into a more innovative society. After all, while patents are considered a measure of technology prowess and innovation, their quality and value can vary widely. We will come back to that later in this chapter.

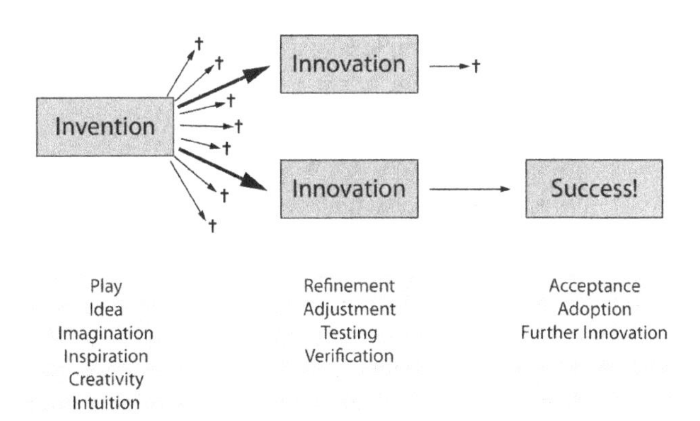

Fig. 1.1. Macro-inventions can arise from an idea or just play and usually require high doses of creativity and imagination. Before they are recognized as macro-inventions, the original inventions need to be optimized through a process called innovation and adopted by society. After that they become subject to endless rounds of further innovation by micro-inventions.

Problems with innovation in Asia were also on the mind of the Korean business mogul Lee Kun-hee when he recently returned to the chairman-

9 Lohr S. "When innovation, too, is made in China", *The New York Times*, January 2, 2011

ship of Samsung, the Korean electronics giant and the world's largest maker of memory chips and flat panel displays. Striking a somewhat apocalyptic tone he argued that Samsung lacks a clear vision on what its next-generation products should be and expressed doubt if the company, whose success is based on rapid manufacturing and iron discipline, can reinvent itself as a creative business[10]. Therefore, it seems that declining innovativeness is not merely an American problem, but something we share with other successful states.

The red flag of an innovation shortfall in the US may be considered as an ominous sign, especially after the great recession of 2008 (in 2009 the US government literally had to bribe the car industry to innovate!), yet does not tell us much about long-term patterns. Indeed, technological progress is often hard to predict, as we have seen for the aforementioned process of horizontal fracking that has now given us new, enormous supplies of natural gas and oil. Inventions are often unexpected events that only in retrospect look inevitable. Coming out of nowhere, they can transform society within a short time. Or for others it can take years to mature. Therefore, to get a clear view of technological progress we need to measure long-term patterns of innovation in an objective way. Making such measurements and plot them against time seems a simple way to find out whether we really do experience a technology slowdown.

The most interesting attempt to measure the rate of innovation is a study by Jonathan Huebner, a physicist working at the Pentagon's Naval Air Warfare Center, the results of which were published in 2005. Huebner used two measures for technological progress. First, he used a book by Bryan Bunch and Alexander Hellemans[11], in which they compiled a list of 8583 'important events in the history of science and technology'. Of this list, Huebner only used the 7198 events from the end of the Middle Ages to the present time. He then defined the rate of innovation as the number of important events per year divided by the world population. The rationale for standardizing the data in this way is based on the assumption that population growth over time also increased the number of inventors and should therefore result in more inventions per unit of time. The results are shown in Figure 2A: the rate of innovation reached a peak in the 19th century and

10 Jung-a S, Oliver C. "Samsung's corporate culture in focus". *Financial Times*, March 24, 2010.

11 Bunch B, Hellemans A. *The History of Science and Technology*. Houghton Mifflin, New York, 2004.

then declined throughout the 20th century[12]. While the absolute number of innovations was greater in the 20th than in the 19th century, the proportional increase in world population was greater still. Huebner's decline in inventiveness, therefore, already starts in the 20th century.

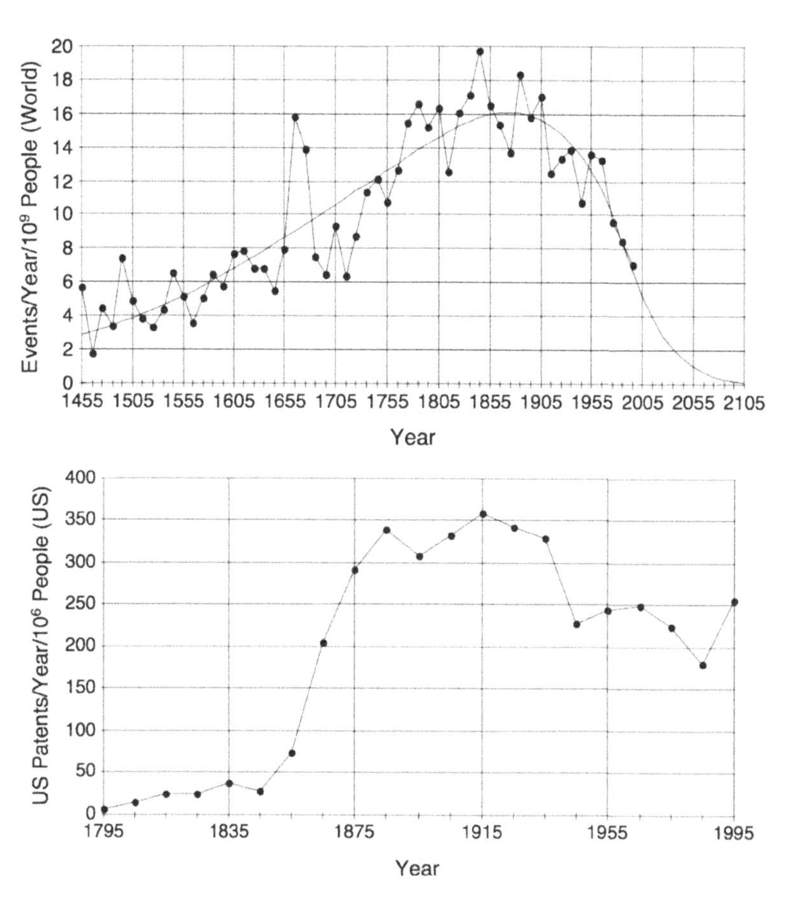

Fig. 1.2. Rate of innovation since the 15th century as calculated by Jonathan Huebner from the list of technology events compiled by Bunch and Hellemans divided by the world population (top) or the number of patents issued each year to US residents divided by the US population (bottom). (Reprinted with permission from Technological Forecasting & Social Change 72 (2005) 980–986.

As a second measure of technological progress, Huebner used the number of patents issued each year to US residents by the US Patent Office di-

12 Huebner J. "A Possible Declining Trend for Worldwide Innovation". *Technological Forecasting & Social Change* 72(8):980-986, 2005.

vided by the US population. The use of patents rather than inventions circumvents the problem that a large fraction of the world population resides in poor countries and is not in a position to contribute much to technology development. Using patents as a metric, the peak in the rate of invention occurred in 1916, 43 years later than when using Bunch and Hellemans' list of inventions. This is shown in Figure 2B.

Based on these results, obtained by using two different measures for technological innovativeness, Huebner concluded that the rate of innovation has definitely peaked and is now rapidly declining. This conclusion, however, seems premature. Before we discuss why that is, there is one thing that should be clear: Huebner's work, published in a reputed scientific journal, is serious stuff. As a scholar, he follows a scientific approach and the article is full of mathematical formulas. But as all scientific work, its results depend entirely on the data input, which was already criticized a few pages further in the same journal, first briefly by Theodore Modis, an expert in strategic forecasting and then by John Smart, a futurologist, in a lengthy and very thoughtful article[13]. I will try to summarize their main criticisms and add some of my own.

First, Huebner's data source for innovations is not well-defined. Bunch and Hellemans really list events that were considered critical for technological progress. They do not distinguish inventions from what merely are modifications or improvements and also count multiple events that really involve the same invention. Huebner realizes the latter, but argues that it is especially for recent years that the same invention or innovation was counted multiple times. He gives the example of the space shuttle missions listed as 37 independent events, but really involving only one innovation or invention. Hence, the decline in innovativeness would be even worse if he would correct for all such marginal improvements and mention only the really important ones. The difficulty of distinguishing mere improvements from real inventions will come back later and is the reason why some scholars advocated using more objective substitutes for technological progress, such as actual research and development expenditures or the number of patents issued. R&D expenditures, of course, tell you little about the actual results in terms of useful inventions. To some extent this is also true for patents,

13 Smart J. "Measuring Innovation in an Accelerating World": Rev. of "A Possible Declining Trend for Worldwide Innovation," Jonathan Huebner, *Technological Forecasting & Social Change*, 72(8):988-995, 2005.

a metric Huebner also used, alongside the number of innovations. While this seems like an objective measure, Smart correctly argues that patents are nowadays mostly issued to corporations, with their value often in litigation.

Smart also criticized Huebner's use of normalizing his data to world population, i.e., dividing the number of inventions by the world population. Such normalization seems to make intuitive sense, but how do we know if inventiveness is really linearly related to population size? For one thing, it needs to take education measures into account, since large numbers of young people on this globe are much less in a position to make breakthrough inventions than the average citizen of England at the time of the industrial revolution. To his credit, Huebner did take this into account and points out that when he would normalize to world per capita GDP or to world per capita education the same decline in innovation would have been observed.

As expected, Huebner's conclusions were vehemently disputed by two major proponents of an exponential increase of technological progress, Ray Kurzweil and Eric Drexler. We will hear more about them later. Their main criticism, similar to Modis and Smart, is the list of over 7000 events in science and technology Huebner used, which they consider arbitrary and not a good basis as a measure of innovativeness. Kurzweil and Drexler believe that developments in their specialty, i.e., artificial intelligence and nanotechnology, respectively, will soon prove Huebner wrong. While this may or may not be true, I note that developments in both artificial intelligence and nanotechnology have thus far failed to fulfill such promises. Both fields are tainted by much exaggeration and hyperbole and it is difficult to see why we should be convinced that there are no obstacles to their eventual success in giving us products that transform our lives. It seems to me that when technological progress is indeed exponential we should be able to see its transformative products all around us.

A very quick way for everyone to see if Huebner could be right about a technological slowdown is to take a look at Wikipedia's *Timeline of historic inventions*[14]. Wikipedia is a free, web-based encyclopedia that anyone can edit. From the 18th century onwards Wikipedia lists the number of inventions per decade. Since Wikipedia can be edited continuously these numbers frequently change. But just to get an idea how the editors (presumably individuals with an interest in technology) perceive technological progress I plotted the number of inventions found in November 2010 in Wikipedia over the period

14 http://en.wikipedia.org/wiki/Timeline_of_invention

1800 to 2010 (Figure 3). The results suggest that innovation reached a plateau late in the 19th century to decline sharply after 1990. Since these numbers are not corrected for population growth it is not surprising that they do not confirm Huebner's conclusions of a technology slowdown already beginning late in the 19th century. However, they certainly do not suggest that the technophiles, who presumably created these lists, perceive technological progress as ever accelerating. Since Wikipedia is written collaboratively by volunteers from all around the world it is difficult to argue that a technology slowdown is merely a subjective conclusion drawn by a few pessimists.

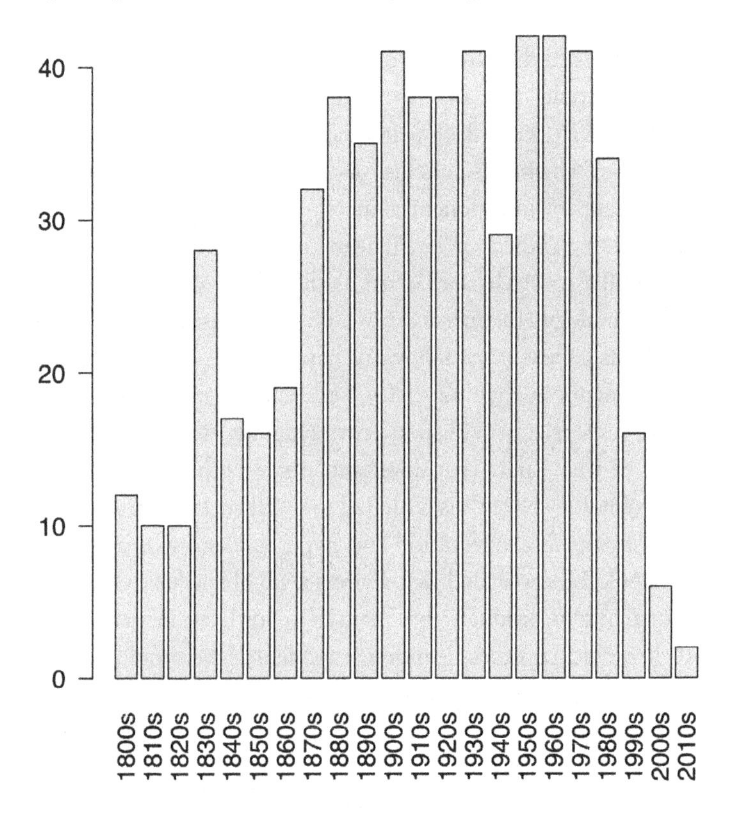

Fig. 1.3. Number of inventions per decade from 1800 to 2010 as taken from Wikipedia's timeline of historic inventions.

Similar to the survey of Bunch and Hellemans that was used by Huebner, Wikipedia's list of major inventions is arbitrary. Moreover, the relative importance of each invention is obvious only to those familiar with the technology and little attention is given to differentiating between major inventions and mere improvements. Of course, one could argue that if technology really progresses ever faster the pattern would be pretty obvious and we should be able to point to numerous major recent inventions.

In his book *The Lever of Riches*, Joel Mokyr introduces the term macro-inventions to define technological breakthroughs that allow novel applications significantly impacting society and changing the way we live and do business[15]. They are distinct from micro-inventions, which are improvements of macro-inventions. Micro-inventions are very important, but merely as logical follow-ups of their critically important brethren in the process of innovation. For each macro-invention there can be thousands of micro-inventions, but sometimes also additional macro-inventions that stand out because they enable applications not allowed by the original macro-invention alone. The best example is agriculture. Invented for the first time more than 10,000 years ago, micro-inventions that improve agricultural output continue to be made even in our own days. However, agriculture also led to macro-inventions, such as irrigation, the seed drill, chemical fertilizer and genetically modified crops. In their time these inventions boosted agricultural productivity so immensely that they deserve to be labeled as macro-inventions. In turn, each of these follow up macro-inventions collected numerous micro-inventions of their own.

Joel Mokyr also recognized that there is an analogy between technological progress and the process of evolution by natural selection[16], which as a biologist and geneticist obviously appeals to me. Also in evolution, as now well recognized, 'progress' occurs in leaps and bounds, with slow gradual change punctuated by the more rapid changes that ultimately gave us new species. The latter happens only once in a while, and only when the environment provides the opportunity for the new species to thrive. While this is an overly simplistic way of explaining the emergence of new species in nature, Mokyr's argument here is that while there are lots of technological changes, it is only rarely that an idea emerges, so powerful that under the right conditions it will be widely applied, thereby significantly altering society.

However, it is not just a radical new idea that typifies a macro-invention. As we have seen, if not adopted the invention will not even be recognized

15 Mokyr J. *The Lever of Riches: Technological Creativity and Economic Progress*. Oxford, New York, 1990.

16 Mokyr J. "Punctuated Equilibria and Technological Progress". *American Economics Review*, 80, 350-354, 1990.

as such. Adoption is obviously much more difficult for a macro-invention than for a micro-invention. Similar to evolutionary change, new technology must find itself in a receptive environment providing it the opportunity to be further refined and penetrate society. If the circumstances are not right, the new idea may end up being just a curiosity. For example, Leonardo da Vinci's design from 1493 of a screw-like device, which according to him would rise into the air if turned at a sufficient speed, is not the same as the helicopter, which was invented early in the 20th century. In the same vein, it can be argued that the aeolipile, a vessel that rotates on its axis, driven by steam projected from nozzles and invented by the first-century mathematician Hero of Alexandria, was a predecessor of the steam engine. However, at the time it was merely a gadget. An actual steam engine, i.e., the first one practically used on a relatively large scale, was invented in 1712 by Thomas Newcomen. But note that Newcomen's invention did not fall out of thin air. Thomas Savery had already patented a first crude steam engine in 1698. And others would argue that it was really James Watt who in the second half of the 18th century made the critical improvements for the steam engine to evolve into the viable piece of machinery that became the heart of the industrial revolution. After this first cluster of inventors there were many others who made improvements, some of them important inventions in their own right, but no longer macro-inventions.

The advantage of Mokyr's concept of macro-inventions is that everyone intuitively knows what they are and what they are not. Agriculture is a macro-invention because for hunter-gatherers deliberate planting and harvesting was a major deviation from regular practice and, being far from obvious, required creative thinking and experimentation. After the first crop, however, planting other crops merely involved further innovation with many micro-inventions. In contrast to agriculture, industry - the production of an economic good (either material or a service) — is not an invention, but continuous practice since times immemorial. Within industry there are plenty of macro-inventions, but unlike agriculture industry itself is not an invention.

A macro-invention is a windmill, but not any of its many improvements that run straight up to our current, powerful wind turbines generating clean electricity. A macro-invention is the mouse and graphical user interface invented by Douglas Engelbart and others, which turned computers from specialized machinery into a user-friendly tool, but not the Apple

Macintosh or Microsoft Windows, both based on Engelbart's invention. However, the Apple's and PC's were the applications that made Engelbart's invention a macro-invention. They were the fertile ground absorbing Engelbart's ideas to let them change society by making computers a household product. Hence, a macro-invention!

Thus, while there are lists of thousands or even tens of thousands of inventions, the number of macro-inventions is conveniently small. I tried to follow Mokyr's definition in putting together a list of macro-inventions from 10,000 BC to the year 2010. Inventions made before 10,000 BC include fire, fishing gear, bow and arrow etc., and stretch back to our origin almost 200,000 years ago. Around 10,000 BC we see the beginning of agriculture and a dramatic increase in inventions. The list counts just over 300 inventions, which makes it easy to verify if I missed major inventions or included mere improvements as genuine macro-inventions. Some of what I included or not included will be open to debate. For example, were the paper clip and staples really macro-inventions and comparable to the internal combustion engine? I think they were, because not only are they still with us in a surprisingly large number of designs and applications, but they have also dominated office environments for over a century. Even now, in spite of our embrace of the paper-less, digital office, we still cannot do without them. I feel that the far majority of what is listed will not be disputed.

Of the more than 300 macro-inventions listed in the table, fewer than 50 were made before 500 BC, with over 100 only from the last 100 years. What this tells us is that inventions accumulated faster in more recent times, which confirms the expectations of most futurologists and is undoubtedly something most of us would intuitively expect. Sometimes intuition is right and when plotting the number of new inventions since 500 BC, we do indeed see an exponential increase (Figure 4A). The same is true when plotting not the accumulated number of inventions, but the number of new inventions per year, i.e., a measure for inventiveness (Figure 4B). However, from this graph we can also see the beginning of a decline around 1970 with no sign of a restoration yet (see also the inset of Figure 4B).

Hence, from whatever way you look at it, simply by measuring what are considered major inventions, either intuitively as by Bunch and Hellemans and in Wikipedia, or following Mokyr's definition of macro-inventions as I tried to do, there is no recent pattern of ever accelerating rate of macro-invention. In fact, everything suggests a dampening of humanity's inventive streak. Before we turn to what may be the cause of this slowdown, some possible sources of error should be considered.

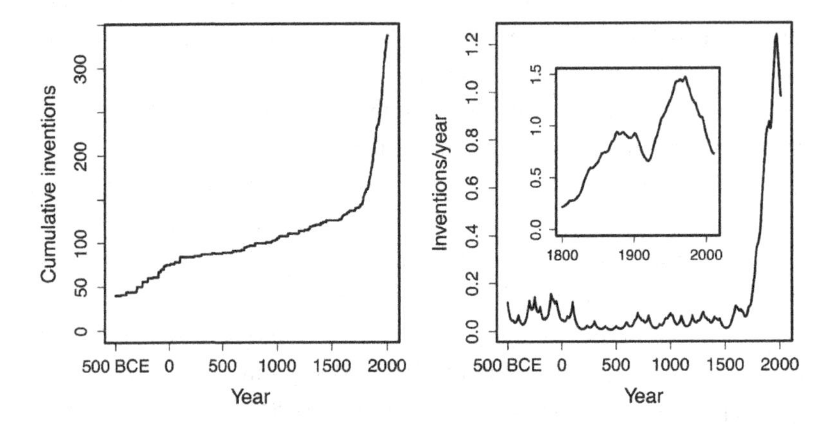

Fig. 1.4. Over the long haul, major inventions made by humankind as a whole accumulated exponentially (A) and the same is true for the number of inventions per year (a measure for inventiveness), until 1970 when inventiveness begins to decline (B).

First, one could perhaps argue that macro-inventions need decades before they are recognized as such, which would explain the rather steep decline in inventiveness after 1970. I think this is very unlikely. Our society is flooded with information about new technology. Perhaps because we are all so convinced that we live in the most inventive of times, no chance is left unused to bring new, potentially breakthrough inventions to the attention of the general public. Entire magazines are devoted to new technology and most newspapers have frequent special sections on the most recent progress in technology. It is therefore difficult to see how major inventions can systematically be overlooked only because their potential usefulness is simply not yet broadly recognized. Indeed, in my own list I gave our own times the benefits of the doubt by listing inventions that have not proved their mettle yet. For example, the scramjet is not yet applied and may never prove useful, the coronary artery stent may never be the revolutionary step forward in preventing heart attacks as some think, and induced pluripotent stem cells, which in theory make it possible to grow replacement cells for any organ from a simple skin biopsy, is too recent an invention to permit a well-founded prediction of its usefulness.

A second possible reason we may underestimate the number of recent macro-inventions is that we miss them because something in the nature of the inventive process has changed over time. An example is so-called end-user or consumer innovation. In his book, *Democratizing Innovation*, Eric von Hippel, a professor of technological innovation at the Massachusetts Institute of Technology's Sloan School of Management, shows how users of technology can themselves become innovators and develop creative new applications around a given product[17]. An example is the social networking product 'Twitter', and many other internet applications. However, while economically important, it is difficult to see how such new uses of a given technology differ from micro-inventions as defined by Mokyr (see above).

The aforementioned John Smart proposed a somewhat similar argument for why macro-inventions may be missed. His point is that we may no longer recognize groundbreaking innovations because there are so many complicated processes associated with a given technology that you need to be an expert to recognize a transformative change. To some extent this may be the explanation of why horizontal fracking was recognized so late as a major invention; much of the innovation that is going on in oil and gas exploration takes place under the surface (literally in this case!) and has become all but invisible to the average person. However, most of such 'underground' innovations are micro-inventions, which almost always go unnoticed by the general public. The occasional event that proves to be a macro-invention only later is unlikely to change the pattern.

Great expectations!

While we struggle with identifying transforming and groundbreaking inventions in our own time, the generations that went immediately ahead of us had no such problem. When visiting St. Mary's Hospital in London in 1991 my host, Dr. John Hardy, one of the world's renowned experts in the genetics of Alzheimer's Disease, draw my attention to a couple of windows on one of the upper floors as the place where Alexander Fleming once had his laboratory. In that exact spot, Fleming discovered penicillin in 1928. Considered a wonder drug, penicillin and the range of other antibiotics following in its track may well be one of the greatest inventions of all time[18]. It saved humankind of the scourge of infectious disease, which killed people

17 von Hippel E. *Democratizing Innovation*. MIT Press, Cambridge, 2005.
18 Luiggi C. "The discovery of penicillin, circa 1928". *The Scientist* 24, 84, 2010.

indiscriminately, from women giving birth and soldiers wounded on the battlefield to those unlucky enough to get pneumonia, tuberculosis or other, often lethal infectious diseases. Nowadays it is simply unimaginable how high infectious disease-related mortality rates once were. Fleming's invention virtually ended this curse, contributing greatly to a dramatic increase in life expectancy which was already under way due to similarly great inventions in agriculture and general hygiene. It also, and this is a good reminder of the power of technology, greatly reduced the cost of healthcare. Imagine, all those tuberculosis patients so vividly described in Thomas Mann's *The Magic Mountain*[19] languishing for years before most of them inevitably found their death. And sanatoriums are costly! Nowadays most of these diseases can be treated at home and they usually clear up in 2 to 3 weeks.

While one can argue that the 19th century was the most innovative era in human history, it was really in the 20th century that the impact of all that technological progress was fully realized. To some extent this is due to the rapid expansion, early in the 20th century, of 19th century innovations to ordinary people (at least, in Europe, the US and Japan), which made technological progress part of everyday life. The industrial age was considered by most people as a great achievement, with all kinds of innovation enthusiastically received by the world. Technological advancement was seen as humanity's manifest destiny and could do no wrong.

According to David Gelernter in his 1995 book *1939, The Lost World of the Fair*[20],

> That was the age of technology par excellence. Technology then dealt in the tangible and the everyday, not in strange stuff like software and silicon. Technology meant refrigerators and radios. It was about new ways to build bridges, dams and skyscrapers, not about space stations and supercolliders. It was about Lucite and Plexiglas, both on display at the fair. It was about nylon stockings — available as of December 1939 — and rayon and cellophane. It was about elevators: Visiting from France, Le Corbusier, ordinarily a fierce critic of all things American, was enchanted by American elevators capable of powering big loads up and down thousand-foot buildings in a flash.

And without a doubt, the 1939 New York World's Fair must have been breathtaking. Its theme was 'The World of Tomorrow', and the moving map of the 'America of Tomorrow' exhibit showed beautiful highways and clo-

19 Mann T. *The Magic Mountain*. Vintage, New York, 1996.

20 Gelernter D. *1939: Lost World of the Fair*, p. 226. Avon Books, New York, 1995.

verleaves with cars carrying people to skyscrapers. Exhibits forecasted the houses and cities of tomorrow and presented streamlined trains, modern furnishings, television, and talking robots.

However, during war time the dark side of technology is unveiled. The great war of 1914 to 1918 had given the world a taste of the horrors of the machine gun, battle tank and fighter plane. That was technological progress also! Perhaps the best representation of technology's Janus head is the 1927 movie *Metropolis* by Fritz Lang, which paints a bleak picture of a capitalist class society, contrasting the gleaming office towers, connected by monorail to living centers and amusement parks, with the subterranean slums where the workers lived. Here, the city of the future is depicted both as a great place to live and an inferno of cruelty. However, the movie ends happily, with the class divide bridged by agreeing to distribute the technological dividend more fairly among rich and poor.

Whether considered a blessing or a danger, in 1939 everybody seemed convinced that technological progress, more revolutionary than ever before, was inevitable and would continue to transform society as dramatically as the earlier inventions of steam engine, automobile and electricity. And after the Second World War, when the US and Europe experienced their golden age of economic growth, technology was still most people's darling, from the architectural designs of utopians such as Richard Buckminster Fuller to the future world societies that sprang from the imagination of the latest crop of science fiction writers.

Science fiction now became much more confident in making precise predictions of technological progress. Indeed, predictions made in the 19th or early 20th century by such fathers of science fiction as Edgar Rice Burroughs, H.G. Wells, Jules Verne and Aldous Huxley were imprecise about technology but the predicted outcomes appeared to come true easily. For example, Jules Verne's moon gun in his 1865 book *From the Earth to the Moon*[21] could not work in reality and his other predictions of future technology lacked scientific detail. Yet, we did go to the moon in 1969!

In his 1863 book, *Paris in the Twentieth Century*, the manuscript of which was only discovered in 1989, Verne made some bold predictions of how Paris would look like in 1960[22]. In the book he describes the Paris Metro,

21 Verne J. *From the Earth to the Moon.* http://jv.gilead.org.il/pg/moon/.

22 Verne J. *Paris in the Twentieth Century: Jules Verne, The Lost Novel.* Ballantine, New York, 1996.

driven by compressed air and running over viaducts, motor cars with gas combustion engines and electric street lights connected by underground wires. Remember that this was written in 1863! The first Paris subway line opened in 1900, the light bulb was invented in 1875 and the first automobile driven by the internal combustion engine was from 1889.

Again, while the bold predictions of these fathers of science fiction usually lacked accuracy, they were astoundingly on target in their essence. After all, we do live in a world with airplanes, helicopters, television, computers, automobiles, space flight, guided missiles, skyscrapers of glass and steel, and submarines carrying people over distances of hundreds of miles below the surface of the sea. While Verne was a lawyer who wrote librettos for operettas in his spare time, and Rice Burroughs a discharged soldier occupied by ranch work, the post World War II writers of science fiction were often scientists themselves. For example, Arthur C. Clarke, a mathematician and physicist, described communication satellites in 1945 in so much detail that he is generally considered as the inventor of the geostationary satellite, and with the 'Newspad' in his novelization of *2001: A Space Odyssey* he quite accurately predicted Apple's iPad[23]. In his 1976 book *Shadrach in the Furnace*, which takes place in 2012, Robert Silverberg described in intricate detail a liver transplantation[24]. In his famous *Foundation Trilogy* of 1951, Isaac Asimov, a biochemist, accurately described a pocket calculator[25].

Work by the post World War II generation of writers is considered an example of the so-called hard science fiction, which is a category of science fiction characterized by scientific and technical detail. And, clearly, some of their detailed predictions did come true. However, unlike Verne and his generation, who produced a rough outline of the world of the 1960s, the SF authors who actually lived in that time and whose job it was to further extrapolate what seemed to be ever faster progress failed almost completely. They had it all wrong with their new, inexhaustible energy packs, smart underground transport and a steady intensification of the exploration and exploitation of space. Even Silverberg's assumption that by 2012 (when *Shadrach in the Furnace* takes place) all international air travel would be supersonic proved wrong. Almost nothing of these grand predictions came true. Apparently, while incomparably more adept in understanding sci-

23 Clarke AC. *2001: A Space Odyssey*. Penguin, New York, 1986

24 Silverberg R. *Shadrach in the Furnace.* University of Nebraska Press, 1976.

25 Asimov I. *The Foundation Trilogy*, Doubleday, 1951

ence and technology than Verne and his contemporary authors, they were not able to predict what society would look like only five decades ahead. Clearly, they did not lack expertise and neither were their expectations unreasonable. Instead, their optimistic forecasts were disappointed due to the dip in technological progress that is the topic of this book.

We are now confronted with what seems like a paradox. On the one hand, careful analysis shows that the rate of technological progress is on the decline, with much less adopted macro-inventions over the last 50 years than we have become used to since the beginning of the 19th century. However, while the technology-minded novelists from the 19th century were easily proved right, the much more rational forecasts made by their scientifically-skilled successors, some as late as the 1970s, have not been realized. Why can we no longer make breakthroughs as great as those of our immediate ancestors and realize the dreams of Isaac Asimov, Robert Silverberg, Arthur C. Clarke and many others?

A level of uncertainty is understandable when dealing with a phenomenon like technology in the context of human society. Unlike physical processes, social and economic phenomena lend themselves poorly to empirical testing. Exceptions never disprove a theory, and variation, complexity and multi-causality are the norm. Therefore, the conclusion of a decline in technological progress does not tell us much when it cannot be explained. But perhaps there are aspects of technological progress we do not understand and that were also overlooked by the early twentieth century generation who made such bold predictions. Finding the reasons why major technological breakthroughs no longer emerge as rapidly as in the past would strengthen the conclusion of a technological slowdown and improve its acceptability. The best way to begin doing that seems to be by taking a closer look at some of the major areas of technology development and how they fared in recent years. This may give us some clue as to what the practical problems are that keep us from the technological breakthroughs that would continue to radically change society.

Chapter 2. Energy: A Marriage of Convenience

If one has to mention a single technological breakthrough that defines modern industrial societies, than that would surely be the conversion of fossil fuels, i.e., coal, oil and natural gas, into mechanical energy, developed in the 19th and 20th century. Now, at the beginning of the 21st century, coal, oil and natural gas remain the major energy carriers. Coal raises almost half of all electricity in the US and most other countries. It is doing that already since the 1880s using a steam turbine to drive an electrical generator, an application invented by the British engineer Charles A. Parsons in 1884. While many improvements have been made, all power stations are essentially based on this same method and coal remains the least expensive power source and the backbone of our society. Oil remains by far the most important transportation fuel and natural gas heats most of our homes and fuels our cooking since I was a little boy.

Surging demands, especially for petroleum and natural gas, the increasing costs of finding and producing new reserves, and growing concerns about atmospheric greenhouse gas concentrations have now led to the realization that alternatives for fossil fuels must be sought. The main challenge to current alternatives is to compete with fossil fuels, which remain more cost effective. This explains why investments in alternatives are minimal. For example, Exxon Mobil, the world's second largest company, recently decided to invest $600 million, over an extended period of time if research and development milestones are met, in developing biofuels made from algae[26]. While this looks like a serious amount of money, it is really only a tiny fraction of Exxon's investments in oil-related research and development. In fact, the annual research and development budget of this oil giant already amounts to $600 million! The same is true for another oil giant, BP, which even changed its name from British Petroleum to Beyond Petroleum to show its intention of going green. Also BP's investments in alternative energy are minimal, with many more dollars spent, for example, on extracting oil from Canada's tar sands. Clearly, energy companies are not eager to abandon their profitable niche of fossil fuels.

Currently, the only immediate alternatives for fossil fuels in generating electricity are wind and solar. Especially wind turbines are now a very serious alternative for coal-fired electricity plants. The earliest-known design of a wind mill used for water pumping or grain grinding is from Persia, probably around 700. Wind mills were greatly improved by the Dutch from the 14th to the 17th century, who used them not only to grind grain but also for creating 'polders', land below sea level enclosed by embankments called dikes as barriers against the sea water, which is continuously pumped out. First advocated by environmentalists in the 1970s as an alternative way of generating electricity, current wind mills (now called wind turbines) for commercial production of electric power can have a capacity of over 7 megawatt (MW) and reach an overall height of 198 m (650 ft). In comparison, a typical coal-fired electricity plant is about 500–1000 megawatts. All together, wind power contributes around 2% of total US electricity production, about the same as the world as a whole. But wind power represented 40% of all new U.S. electric generation capacity in 2009.

When we talk about solar power, we usually mean photovoltaics, which is a method of generating electrical power by converting solar radiation into direct current electricity using semiconductors. The photovoltaic effect was already described in 1876 by Adams and Day in solid selenium. However, the first practical solar cells were only described in 1954 by scientists at AT&T (Bell Labs) and RCA (RCA Labs). Photovoltaic cells are exceedingly popular and now contribute to electricity production by powering traffic lights and lights for buoys at sea, but also homes and businesses. Large parks of solar cells now have a capacity of about 100 megawatt. Their contribution to overall electricity production, however, is low and doesn't exceed 1%

26 Howell K. "ExxonMobil Bets $600 Million on Algae". *Scientific American*, July 14, 2009.

worldwide. Another form of solar energy is solar thermal, which generates electricity the old-fashioned way by heating water and generating steam, but in this case using mirrors or lenses as collectors to concentrate sunlight as the heating source. The heat can be stored for 7 to 8 hours, for example by using molten salt[27]. In the US there is only one solar thermal reactor (in California) but more are being planned.

Theoretically, there is no reason why wind and solar could not substitute for fossil fuels in generating electrical power. They would not make electricity any cheaper, but are certainly a viable alternative for fossil fuels. In fact, it has been calculated that a network of land-based 2.5 megawatt wind turbines restricted to non-forested, ice-free, non-urban areas operating at 20% of their rated capacity could supply more than 40 times current worldwide consumption of electricity[28]. However, it is obvious that both wind and sunlight are highly distributed energy sources and also not always available. That is, they require much space and are only economical in places with a lot of wind or sunshine. Those are rarely the places where most of our electricity is used. Because electricity is difficult to store, it will be necessary to drastically reconfigure the electric grid so that electrical energy can be moved from empty places with a lot of wind or sun to urban areas. Integration of the grid over large distances, such as all over North America, Europe or China would allow moving power from places where it is generated to the cities and compensate temporary local shortages[29]. After all, the wind is always blowing somewhere.

Electricity is generally less suitable for fueling transportation because of a lack of options to store electrical energy. Indeed, the low capacity of batteries is a major frustration for cell phone and lap top users, as well as the main reason for the premature death of General Motors' EV1 electric car (rather than a conspiracy by the auto and oil industries). Battery development has progressed very slowly, but with the invention of the lithium ion battery in 1991 and the innovative use of these batteries in the first commercially available all-electric car by Tesla, a start-up company in California, our options to leave petroleum behind have now become very good. As de-

27 Roeb M., Müller-Steinhagen H. "Concentrating on solar electricity and fuels". *Science* 329, 773-774, 2010.

28 Lu X, McElroy MB, Kiviluoma J. "Global potential for wind-generated electricity". *Proc. Natl. Acad. Sci. USA* 106, 10933-10938, 2009.

29 "Where the wind blows". *The Economist*, July 28, 2007.

scribed in the next chapter, the first electric cars equipped with such batteries are already on the market.

Always 50 years away!

Altogether, wind and solar are hardly breakthrough inventions and it has taken decades of painful optimization before we reached the current stage of practical usefulness. Even now the large-scale implementation of wind and solar to replace coal and natural gas in generating electricity will take time and some serious upfront investment. While there are other alternatives, most notably biofuels, none would revolutionize the way we currently generate energy or make it any cheaper or more efficient as we surely all expected half a century ago. At that time a logical follow-up of fossil fuels was generally considered to be atomic energy, or so it appeared in the 1950s to many of the scientific community and the lay public alike. In the aforementioned science fiction classic *Foundation Trilogy*, written by Isaac Asimov in 1951, our future homes, vehicles and space ships were all powered by atomic energy. Moreover, atomic power sources were also miniaturized and capable of driving simple, portable tools, such as an atomic shear for cutting steel. And, of course, ways had been found to make atomic energy harmless!

While obviously a work of fiction, offering no detail as to how some key problems would have been resolved, Asimov also wrote a great amount of nonfiction, most notably popularized science books explaining basic scientific concepts. Indeed, there were undoubtedly a great many professional scientists and engineers in the 1950s who would have predicted that at the turn of the century atomic energy would have all but eclipsed coal as a power source, providing cheap and plentiful electricity.

Nuclear power, developed during the 1940s and 1950s, actually has a considerable share in our electric power generation (in some countries such as France it is even the main source of electricity). Like a coal-burning power plant, a nuclear power plant operates by heating water into pressurized steam, which drives a turbine generator. The difference is that it generates its heat by splitting uranium atoms (called fission), which generates no carbon dioxide or other heat-trapping gases. Instead, it leaves behind highly radioactive waste, and nuclear power plants can also be dangerous as previous accidents have shown, most notably the one in Chernobyl, Ukraine, and the 2011 calamities in Japan.

Hence, while nuclear fission is a green power source, the safety risks and the radioactive waste problem increase the costs of building and operating a power plant based on this principle. New, improved designs with passive safety features that do not require a power source to shut the system off after a mishap are available for decades. Yet, most developed countries, including the US, Italy and Germany, began to abandon it. More recently, as a consequence of the realization that fossil fuels are ultimately limited and cause greenhouse gases, nuclear energy is again promoted as the only clean alternative to coal. Nevertheless, it seems doubtful that this source of electricity, irrespective of the greatly improved safety measures will ever be able to compete with coal or natural gas. This is due to increased public resistance.

The situation today is vastly different from the 1960s when most nuclear power plants were planned and built. Proposals for re-opening long shuttered uranium mines and building new processing plants and reactors are greatly hampered by public resistance. Whether or not the industry is right with its claims that we have learned from past mistakes that led to accidents, cancer deaths and environmental destruction, and that safety has greatly improved, the 'not in my backyard' (NIMBY) effect is very strong. And backyards nowadays can stretch awfully far. Aggressive investing would undoubtedly greatly improve safety and efficiency of nuclear energy through further innovation, but that no longer matters. Current attempts to accelerate procedures for licensing a new reactor by applying for site approval early, illustrate the industry's uphill battle. Seeking early approval is driven by the realization that once construction is complete an endless wait for approval to actually run the plant is not good for the bottom line. However, even the right to apply for such pre-approval is now fought in court.

The first generation of nuclear power plants, similar to the first practically used steam engine design by Thomas Newcomen (1712), was really not more than transition technology. However, unlike the Newcomen steam engine, which was succeeded half a century later by James Watt's greatly improved design, further development of nuclear power was stopped in its tracks by increasing resistance from the public, deterring aggressive, further commercial investments. While new designs abound, further progress in this area is delayed and may even be permanently halted because of the fear of accidents. Ironically, if their development would not have been aborted most of the original nuclear reactors could by now have been replaced by much safer models.

Probably the safest type of nuclear energy is nuclear fusion, another type of nuclear power generation. Unfortunately, most people do not know the difference between a fusion and a fission reactor and are therefore conveniently against both. Nuclear fusion is much more powerful than nuclear fission. It uses hydrogen from water as fuel rather than finite supplies of uranium that need to be mined. There is no possibility of a catastrophic accident in a fusion reactor resulting in major release of radioactivity. Although there is some potential exposure to radioactivity and also some radioactive waste products, this is a far cry from fission reactors, whose waste remains radioactive for thousands of years. While a simple idea, i.e., using the energy released when two heavy hydrogen atoms are fused into helium (similar to how the sun generates its heat) to heat water and drive a steam turbine, nuclear fusion proved difficult to achieve in practice. This is due to the difficulty in maintaining the hydrogen long enough at the necessary high pressure and temperature (creating a state called a plasma) to generate a self-sustained fusion reaction that yields energy.

Two major public projects, one in the US and one, international project, in Europe, are currently attempting to harness fusion power for electricity generation. The US project is carried out at the National Ignition Facility (NIF) at Lawrence Livermore National Laboratory (LLNL) in California[30]. By focusing the intense energy of 192 giant laser beams on a tiny (about 5 mm) pellet filled with hydrogen fuel they attempt to create the necessary high temperature and pressure long enough for a self-sustaining fusion reaction. The second, international project is altogether different. The International Thermonuclear Experimental Reactor (ITER) which is being built at Cadarache, France[31], is a so-called Tokamak. Invented in the 1950s by Soviet physicists, a Tokamak is a machine that uses a donut-shaped magnetic field to confine the fuel pellet at the very high temperatures needed to create the plasma and produce controlled thermonuclear fusion power.

The repeated disappointments in getting this process to work since the 1950s have now led many to believe that nuclear fusion is a lost case[32]. But this conclusion is premature and not based on scientific evidence. For example,

30 https://lasers.llnl.gov/

31 http://www.iter.org/

32 Seife C. *Sun in a Bottle*. Viking, New York, 2008; Brumfiel G. "Just around the corner". *Nature* 436, 318-320, 2005.

the endless delays in designing and building an experimental fusion reactor that actually produces energy are merely political. It took a long time before agreement was reached about Cadarache, France as the site for ITER. And then there were all the cost sharing issues. While the costs are high, nuclear fusion could be for the 21st century what the steam engine was for the 19th. Compare, for example, the ITER budget of 16 billion over the 14 years of the experiment with Exxon Mobil's $120 billion on capital and exploration projects between 2008 and 2012. And that is only one of many oil companies.

While the steam era was characterized by individual entrepreneurs, such as Savery, Newcomen, Watt and Parsons, nuclear fusion is dominated by governments and government-paid scientists. To some extent this may explain some unpractical aspects of fusion power that might have been avoided when its development would be in the hands of individual entrepreneurs. While fusion power offers the potential of environmentally benign, widely applicable and essentially inexhaustible electricity, the process is still based on the generation of heat to drive a steam turbine. This is not exactly what Asimov had in mind when promoting atomic energy as the fuel of the future. What he thought of is probably something like a fuel cell, which produces electricity by converting hydrogen and oxygen into water without any other by-products. Fuel cells look like batteries from which they differ by the need to replenish reactant, i.e., hydrogen, while batteries store electrical energy chemically in a closed system. Fuel cells were already invented in 1839 by William Grove, but never caught on because they are more expensive and much less efficient than combustion-based systems, such as traditional engines. Could nuclear fusion ever be miniaturized into a fuel cell-type of reactor, but then way more powerful?

The desire to make fusion reactors simpler and smaller explains the initial enthusiasm for cold fusion, based on claims made by electrochemists Stanley Pons and Martin Fleischmann in 1989 that fusion reactions can be produced at room temperature in a simple cell with a solution and electrodes. These claims have never been confirmed and cold fusion has been largely dismissed by the scientific community. A more recent alternative, however, may be more promising. *Focus Fusion* refers to a so-called 'Dense Plasma Focus' device that acts as a nuclear fusion generator that produces, by electromagnetic acceleration and compression, a short-lived plasma from hydrogen and boron (another common element)[33]. Such a reactor would be

33 http://focusfusion.org/

much simpler and smaller than a Tokamak. Moreover, the energy generated by such a reactor would be released through the motion of charged particles that could be converted directly to electricity, which is distinctly different from Tokamak or laser-based ignition. No need for Parsons' age old steam turbine anymore!

While we are obviously very far away from Asimov's miniaturized nuclear power generators, his expectations were far from unrealistic. Indeed, if our big energy companies would be as entrepreneurial as they were in the past we would likely be much further along. While in the past it never took long for industry to recognize the potential of new inventions for future profits, today they seem to be solidly wedded to fossil fuels. While increasingly strong voices from both governments and an environmentally conscious public now call for investment in other sources of energy, the fossil fuel companies continue to spend only a token fraction of their budgets on alternative energy. Virtually their entire research and development budget is devoted to further improvements in unlocking new resources of oil and gas, either from areas previously closed for exploration, such as nature reserves, from sources previously too costly to explore, or by making improvements in exploiting current sources. Similarly, Big Coal wants to continue its destructive practice of mountaintop mining rather than investing in the future.

The conservatism of our major energy companies seems highly surprising in view of the adventurism that characterized their predecessors early in the 20th century. After all, in those days, someone must have taken the risky decision to invest in coal, oil and natural gas, someone must have decided that electricity had the future and someone must have invested in the infrastructure for transporting large amounts of liquid and gaseous fuels. Why is there no one who tries to get a head start on the others by investing in something that surely must at some point in time replace the limited fossil fuels? One explanation is the enormous costs involved. It is one thing to invest millions in technology (wind, solar, biofuels) that may lack advantages over fossil fuels but is at least green, but quite another to invest billions in a risky bet on nuclear fusion. This is especially true since fossil fuel supplies will last until well into this century and probably longer.

According to Exxon Mobil, by 2030 hydrocarbons — including oil, gas and coal — will still account for 80% of the world's energy supplies[34]. A large part of this will come from North America. Already endowed with

34 http://www.exxonmobil.com/Corporate/files/news_pub_eo_2009.pdf

enormous supplies of coal and natural gas, the US and Canada are also major oil producers. Since 2009 US domestic oil production is again on the increase, with potentially huge supplies in the Gulf of Mexico and the Atlantic coast (still closed for exploratory drilling) untapped. Most notably, the already discussed new methods of hydraulic fracturing and horizontal drilling now lead to rapidly increasing production of natural gas and oil from gigantic shale fields in North Dakota and several other states.

All in all, these bountiful supplies of fossil fuels do not provide great incentives for the industry to change their ways. Current energy giants are in the fossil fuel business since they were born, early in the 20th century. In those early days they might have moved in directions other than fossil fuels alone. Now, however, they are entrenched in the business and built up highly stable organizations with practices accepted by substantial segments of society. They maintain powerful lobby groups to defend their interest with the government and see no reason to heavily invest in alternatives that are either immature, require extensive alterations in societal infrastructure or are simply pies in the sky. They also know that most people, in spite of their occasionally expressed concern for the environment or fear of future price increases, are ultimately satisfied with the status quo. In fact, serious resistance by action groups against some green energy alternatives are not a remote possibility, as tenacious opposition to proposed wind turbine parks has shown[35].

In earlier times, public resistance against technological development was hardly an issue in decisions to invest. Both governments and the media did not have much patience with resisters. This is exemplified by an article in the *New York Times* of July 7, 1890, describing the opposition to the early railroads. The author had nothing good to say about these 'opponents of progress' and scornfully dismisses their arguments as superstition. Compare that to the attention modern opponents of new technology often get! Earlier still, in England, the Luddites, a social movement of British textile artisans who protested against the changes produced by the Industrial Revolution, were executed or transported as prisoners to Australia. Nowadays, public opinion counts, even if it is unreasonable or irrational, and has become part of the business models that dictate investment strategies.

In conclusion, major breakthroughs in the energy sector remain forthcoming with very few macro-inventions related to alternative energy since

35 "Not on my beach, please". *The Economist*, August 21, 2010.

nuclear fission technology, now more than half a century ago (1942). The most logical explanation for this lack of major progress appears to be the continuous focus of the industry on fossil fuels. In turn this is due to a preference for established practice with proven results rather than exposing their companies to excessive, unnecessary risk. Indeed, public disapproval of nuclear energy makes it very difficult to seriously increase investments in that area. In turn, this has essentially prevented nuclear energy from maturing by undergoing a process of gradual improvements through micro-invention similar to what has happened with the steam turbine since 1884.

While energy as an area of technology development has seen no major breakthroughs for more than half a century, lots of progress has been made. Hundreds if not thousands of micro-inventions greatly improved fossil fuel discovery and exploration, optimized its processing and greatly increased efficiency of fossil fuel use. Some of these micro-inventions, such as horizontal fracking, are macro-inventions in their own right.

At least one other energy-related macro-invention is now proving to have a positive impact on human well-being: the solar cell. Solar cells, as we have seen, are now used more and more frequently to generate electric power, mostly connected to the grid. However, solar cells proved to be useful off-grid as well, for example, in powering lights in buoys, traffic signals or lights in your back yard. Such off-grid applications of solar cells now begin to greatly improve the lives of the rural poor in developing countries. For example, most African villages are not connected to the grid. Light at night can only be obtained through kerosene lamps. Using high-efficiency LED lights, another macro-invention (1993), hooked up to a $100 solar system, off-grid electricity has become available to millions of families. Interestingly, the same device not only gives them light at night and their children the opportunity to study, but also allows charging cell phones[36], another macro-invention (1973). Cell phone use has exploded in rural Africa, where many services common in the developed world are often lacking. This may be far from Asimov's miniaturized atomic power packs, but it nevertheless makes a big difference for lots of people all over the world.

36 Rosenthal E. "African Huts Far From the Grid Glow With Renewable Power." *The New York Times*, December 24, 2010.

Chapter 3. Transportation: Fast Forward Interrupted

The fastest and most convenient way of transportation on our planet used to be by ship. Today virtually all goods traded worldwide still travel by sea, but for passenger transport shipping is limited to ferries and vacation cruises. History has seen some major innovations in ship design and ship building, most notably the eclipse of sailing-ships by coal-powered steam boats in the 19th century. Since then, nothing as dramatic has happened and only three major new ship designs were invented: the submarine, the hydrofoil and the hovercraft.

Instead of new designs, shipping has sought progress mainly in relatively modest changes of more of the same. Coal has been replaced by diesel oil and a variety of improvements has made shipping a lot more economical and convenient, but even modern ships would be easily recognizable by someone from the 19th century. Improvements have been made in size and speed as long as they were practical and cost-efficient.

In its days the Titanic, which sank on its maiden voyage in the year after it was launched, i.e., 1912, was the largest ship ever built. It weighed about 45,000 tons and was almost 500 feet long. Today, the largest passenger ship in the world is the Royal Caribbean's *Allure of the Seas*, of 225,000 tons. The ship is 1187 feet long (362 meters) and 184 feet wide (56 meters). It still is beaten by Jules Verne's Leviathan II, moored in the Paris of 1960 in his aforementioned book *Paris in the Twentieth Century*. Verne describes this cruise ship as 1200 meters long and 61 meters wide. However, these days, cargo vessels are the largest ships around, and we have oil tankers of over half a million tons and more than 1,500 feet long. The largest container ships carry as many as 10,000 container boxes across the Pacific Ocean.

Speed has also significantly improved. The clipper was a very fast sailing ship of the 19th century and the fastest cargo ship in the world, capable of speeds of about 16 knots (30 km/h). Nowadays, the fastest ships can reach speeds of well over 33 knots (61 km/h). The longer and thinner a ship is, the faster it will go but at the cost of stability. As known since time immemorial by the ancient people of Oceania, catamarans are faster than monohulls and much more stable. Nowadays multihull vessels from catamarans to trimarans and even pentamarans are used as cargo ships and as ferries. They can reach a maximum speed of around 85 km/h.

While ships are still the most efficient way of transporting goods, humans now move around mainly by motor car, airplane and train. Trains, initially the symbol of the industrial revolution, have significantly lost in popularity since they reached their zenith in the early 1900s. Therefore, it is not the trains that have become the cause of the biggest change in the landscape over the last hundred years, but the network of expressways and the motorcars that are on it. This is truly revolutionary and allows everybody but the very poor to travel comfortably to almost any destination on earth.

The first automobiles as we now know them were introduced in 1907 with Ford's Model T. While today they look very different, their essence has not changed. Indeed, while micro-inventions, such as electronic fuel injection (to replace the carburetor) and the airbag (to increase safety), greatly improved motor cars, the concept of Ford's car for the masses proved very resistant to change. And then I am not talking about attempts to get people out of their cars into public transport, or about the delays in replacing gasoline cars by electric cars. As discussed earlier, inexpensive and sheer unlimited supplies of gasoline has thus far precluded serious investments in increasing battery life or the creation of a new infrastructure for electric cars. But even if we would have electric cars that can be re-charged in a matter of minutes and have a similar driving range as current gasoline cars, not a lot would change. Indeed, while remarkable in itself, such a switch away from gasoline towards a more environmentally friendly power source would not dramatically change our life. That would require a major change in the way transportation is organized. Public transport continues to suffer from the same deficiencies it always had: it brings you to places you don't need to be, at the wrong time, and with a lot of people you don't want to be with. Mass public transit suffers from fixed timetables with often long waiting times, restricted routes and shared travel space. There is no shortage of science writers and science fiction authors making confident predictions of how public and private transport would look like in the future. While they vary, most of them foresaw some sort of tunnel-based network of private vehicles operated automatically.

For example, in the 2004 action movie *I, Robot*, Will Smith plays detective Del Spooner, who has to investigate the apparent suicide of the scientist behind the large robot population which by that time (2035) had become indispensable in serving humanity. (The movie is not really based on Isaac Asimov's novel of the same title, except for the famous 'three laws of robotics' that should protect humans from potential robot aggression.) In the car attack scene detective Spooner is found sleeping in his car, going 280 miles per hour in a tunnel under Chicago, obviously on automatic pilot, when he is attacked by the robots. Apparently, as late as 2004 it was still anticipated that three decades later smart transportation would have become a reality, something the average moviegoer of today may find doubtful. Still, are these auguries that much bolder than the prediction of our current network of expressways made early in the twentieth century?

As I said, predictions of highly automated public and private transport are common in science fiction. I remember the highway scene in *Solaris*, the 1972 Russian movie directed by Andrei Tarkovsky (after the 1961 book by the same title, written by Stanislaw Lem). Tarkovsky's stylized vision of a hands-free car ride through Tokyo symbolizes confidence in future city life as much as anything else. The idea of self-driving cars, such as personal monorail sky cars, is hardly limited to science fiction novels and has been around for years, also in non-fiction environments. To just get into your car and simply type in your destination, and spend time on your emails instead of navigating the streets on the way to work, would obviously appeal to most of us. (Not all of us; consider for example the way most Europeans stubbornly cling to the manual transmission!)

Here and there, we can taste from a future transportation utopia. In 2009, Heathrow Airport in London unveiled its personal rapid transport system in the form of driverless so-called pod vehicles, shuttling people between car parks and Terminal 5[37]. They show up on demand in less than a minute, can be programmed to go to any destination, and accommodate only you or the group you are with, just like in the science fiction novels. The only difference is that this system is highly unlikely to extend its reach to the city of London to combat congestion. Indeed, nowadays such city-wide schemes are deemed too expensive and will probably never materialize. This raises the question as to how our current world-wide system of expressways, which also required huge upfront investments, was realized.

Expressways were really invented in Italy, the land of Marinetti's racing motorcar, where a prototype in the form of the 80-mile *autostrada* from Milan to Varese had been built as early as 1924. However, the first extensive motorway system was built in the 1930s in Germany. The first section was completed in 1932 (a year before Hitler came to power) and just before the beginning of the Second World War about 5,000 kilometers of motorway was completed, going from city to city by picturesque routes with bridges and service stations designed by architects. The Autobahn may have been built based on military considerations, increasing employment, or just for propaganda. A devotee of modern technology with a special fondness for the automobile, Hitler himself (who commissioned the design of the Volkswagen Beetle) declared the nation's future highways a yardstick to measure its prosperity. Nevertheless, the project had begun before Hitler came to power, and Germany was clearly not the only country planning a modern, limited access, divided-lane road system. It was a time when big projects were readily taken on.

Indeed, after Germany the US got its Eisenhower interstate system in 1956, which came at a cost of $119 billion (close to $500 billion today). Construction of the Interstate System did not contribute to the federal deficit because it was financed with revenue from highway user taxes through the Highway Trust Fund established under the Federal-Aid Highway Act of 1956. It is difficult to overestimate the impact of the more than 40,000 miles long system. It dictates daily life for many of us and has become an essential part of society and a cultural icon as well. And this is now true for the entire globe. In the 1970s virtually every country had its share of a global motorway system. To upgrade this system to a hands-free, automated highway network should come naturally and seems to require significantly less of an

37 Danigelis A. "A Pod Car of One's Own". *Discovery News*, September 22, 2010 (http://news.discovery.com/tech/a-pod-car-of-ones-own.html).

investment. Alas, in 2008 we still drive with both our hands on the wheel, need to fill up frequently at a gas station and be prepared for the occasional mechanical mishaps. Is the vision of controlled car rides in the future so unreasonable?

Of course, designs for smart highways to increase safety and resolve the traffic jams have been made a long time ago[38]. Electronic sensors embedded in the pavement were by now supposed to guide autopiloted cars and trucks along highways, maintaining safe distances between vehicles and keep them from drifting sideways. Exactly as Tarkovsky saw it in his movie, vehicles should by now automatically adjust their speed and positions relative to others for safe braking, passing, or lane changing.

Moreover, our cars should no longer be reliant on petroleum as their main source of energy. Electric cars already exist since the 19th century, but with the introduction of gasoline-powered cars they ceased for the time being to be a viable commercial product, due to their limited range, lack of horsepower and the ready availability of gasoline at low price. But, of course, already since the 1970s there is no reason to hold the further development of electric cars hostage to its past, only because we seem to be married to fossil fuels. Fortunately, the situation has changed and there are still entrepreneurs to show that a divorce from fossil fuels is possible, even if this flies in the face of the naysayers.

Elon Musk, who was born in South Africa but lived and studied in Canada and the US, is an engineer and entrepreneur who made his fortune as one of the founders of PayPal, an internet financial services company. Musk co-founded Tesla Motors in 2004 and is its chairman of the board and product architect. His vision is to deliver affordable electric vehicles to mass-market consumers. To overcome the low energy density of nickel-metal hydride and lead acid batteries, Tesla focused from the beginning on lithium ion batteries, which can store much more electricity and hold their charge must better. Due to their high energy density, lithium ion batteries have become the technology of choice for portable electronics and cell phones. With its Roadster of 2008, Tesla produced the first production automobile to use lithium-ion battery cells, which gave it a range greater than 200 miles per charge.

Of course, one could argue, as most everybody did, that to make a good-looking sports car for over $100,000 is one thing, but to make lithium ion

38 Rillings J. "Automated Highways". *Scientific American*, October, 80-86, 1997.

technology affordable for the average driver is something entirely different. What is overlooked, however, is that almost every invention has begun in exactly the same way. It was only with Henry Ford's Model T that passenger cars came within reach of the average Joe and began to generate the huge automotive market we have now, changing our society dramatically. And as we will see, the same was true for airplanes. Meanwhile, Tesla decided to go full speed ahead and produce a much more affordable, albeit still expensive, 5-passenger electric vehicle by 2012[39].

Musk's example of simply doing what seemed economically impossible subsequently set the stage for major investments by all car companies in electric cars. This development was undoubtedly helped by the very high gasoline prices, which reached an all time high of $4 a gallon in 2008. As it looks now, electrically powered cars are on their way to slowly replace their fossil-fuel based counterparts, with gradual improvements in manufacturing, durability and range of batteries and an emerging infrastructure of charging stations. This most likely would eventually have happened without Elon Musk. Yet, it was Musk who spent a lot of his own money on lithium ion batteries to drive a car, against all current business practice. His irrational decision forced the automotive industry forward and now we see that electric drive trains are practical or will be soon enough. It is still the courageous entrepreneur who drives technological progress, and it is no coincidence that in 2002 Musk also founded Space Exploration Technologies (SpaceX), of which we will hear more later in this chapter. Unfortunately, Musk is a rare breed and similar entrepreneurs are now mostly found in information technology making new computer games or financial software, deemed harmless by the public and mostly outside the range of the regulatory machinery.

Because they have no exhaust gases, electric cars bring the creation of transportation networks of tunnels to serve inner cities infinitely closer. The problem here is the high costs of tunnel boring. To address this issue we need integrated boring machines, in which the entire process of tunnel construction is automated, from the boring to the electrical and ventilation systems. We will read more about that in the next chapter.

The obvious question that now comes up is what takes us so long? Why isn't the future now? Evidently, it is not a lack of imagination or inventive-

39 Miller CC. "Tesla electric cars: revved up, but far to go". *The New York Times*, July 24, 2010.

ness that holds us up. More likely, it is reluctance to deviate too much from accepted patterns in society that keep us from the bold moves our ancestors were not hesitant to make in their time. To switch from horse-drawn coaches to trains and motor cars must have been both more expensive (on the short term) and an infinitely bigger step than investing in electric cars, boring tunnels and overlaying the current highway systems with electronic guidance. But, then, while often complaining about air pollution, congested inner cities or traffic jams, most of us are quite satisfied with the current system, which is more luxurious than almost anybody a century ago could have ever dreamt up. And at that time we did not have a mature car industry.

Up in the air

Flying has always been humanity's dream and a sign of power as well as hubris. This is illustrated by the ancient Greek myth of Icarus. In an attempt to escape from Crete with his famous inventor father Daedalus, who constructed wings from feathers and wax, Icarus ignored instructions not to fly too close to the sun. The melting wax then caused him to fall to his death. An airplane in flight, even now, instills beauty and a sense of awe to an observer. This was certainly true in the 1960s, when I collected airplane pictures, visualizing progress from the first, simple propeller planes to the humongous jetliners and sleek fighter jets of the 1970s. Little could I know that the year I decided I was too old for this, the continuous metamorphosis that brought us from the Wright Flyer I in 1903 to the Boeing 747 and Concorde had already come to an end. When my son was growing up in the 1990s, there was little reason to maintain a picture album to record progress in the aircraft industry.

Since its inception early in the 20th century by the Wright brothers, controlled, powered and sustained heavier-than-air human flight became a fairly routine way of transportation for the well-off already before the Second World War. As early as 1913, the first serious passenger plane appeared in Russia, the Sikorsky Ilya Muromets, which could accommodate 14 passengers. Similarly, the US had its Farman F.60 Goliath by 1919. In the 1920s and 1930s, passenger transport became routine, with planes such as the Dutch Fokker F.VII and the US Ford Trimotor and Douglas DC-2, all accommodating 10–15 passengers. A new standard in terms of speed and range was set by the 30-passenger Douglas DC-3, which became the workhorse of the late 1930s and 1940s.

A good example showing that the sky was hardly the limit in those early days of flying is the Hughes H-4 Hercules, also called 'Spruce Goose'. This giant seaplane was a prototype heavy transport aircraft, designed to transport war materiel overseas during the Second World War. Remarkably, entirely made of wood to save aluminum, which was scarce in wartime, it is still the largest aircraft ever built, with a 320-feet wingspan (larger than a foot ball field). Its maiden flight (which would also be its last one) was by the man himself: Howard Hughes. Glorified in the 2004 movie *Aviator* (by Martin Scorsese, with Leonardo DiCaprio playing Hughes), this legendary tycoon had put all his fanatical energy in the program and did manage to bring it to completion. Eventually the program was discontinued because the war was over and there was no need for big seaplanes. Nevertheless, building such a plane that was actually flyable was a fantastic accomplishment and this was more than 66 years ago!

After the Second World War, Lockheed developed the Constellation and Super Constellation (1950). The latter, often considered the most beautiful airliner ever built, could accommodate almost 100 passengers, a little less than the Douglas DC-7, which emerged in 1953. These were still all propeller planes with a cruising speed of at most 350 miles per hour (DC-7). However, the jet engine was already invented in the 1930s in England and Germany. Propeller planes were powered by piston engines. Jet engines sucked in air at the front with a fan, compress the air, mix it with fuel and light it with a spark. The plane is propelled in a forward direction as the result of the burning gases expanding and blasting out from the rear of the engine through a nozzle. This allows jet planes to fly at much greater speeds, i.e., at cruising speeds of about 600 miles per hour, thus cutting down travel time.

The first passenger jet was the de Havilland Comet of 1951, with space for over 40 passengers (over 70 for the Comet 3 in 1952) and a cruising speed of 480 miles per hour. Others would follow suit with the French Sud Aviation Caravelle (1955; 80 passengers), Russian Tupolev Tu-104 (1956; 50 passengers) and the US Boeing 707 (1957; 180 passengers). At the end of the 1950s, true intercontinental and transcontinental non-stop flights were as routine as they are now.

After the 1950s, continuous small improvements drove down the costs of flying significantly and air travel soon came within reach of even the lowest income families. But there were no longer major changes in speed, range or capacity. Indeed, now, in the first decade of the 21st century our airplanes are not any faster or much bigger than they were in the 1960s and the prospects of new, Concorde-type airplanes to avoid 14-hour flights seem dim. Indeed, air travel often is positively inconvenient with delays rampant and lost luggage so common that it has become impossible to check a bag when you will actually need it immediately upon arrival.

This is not to say there has been no innovation in air travel. For example, improvements in design of wings and fuselage have greatly reduced air drag. The so-called supercritical wing and the now familiar 'winglets' attached to the tips of the wings have greatly increased fuel efficiency. Similar to motorcars, airplanes are now more reliable than ever, with very few accidents, noise levels down and autopilots so good that crews have to be kept awake. (In 2009, two Northwest Airlines pilots overflew their Minneapolis destination by 150 miles without noticing.) An important recent development is the gradual replacement of metal by lightweight, composite material, which reduces fuel consumption significantly and allows higher humidity levels, and improvements in engines, making them both more fuel efficient and less noisy. Nevertheless, no improvements that are even remotely comparable to the change from, say, the Douglas DC-3 (the propeller-driven aircraft that in the 1930s and 1940s revolutionized air transport) to the colossal Boeing 747, which first flew commercially in 1970, have been made. Why not?

Lack of progress in air travel with increasing delays and discomfort is not something we anticipated several decades ago. In fact, in the 1960s, the Concorde was only one of several so-called SST (supersonic transport) designs. Only one other, the Tupolev Tu-144, ever flew commercially (if one can use that term for the former Soviet Union). Designs made by Lockheed and Boeing were never realized because of a general protest against Concorde, which was considered too noisy. The US Senate closed down the SST program completely in 1971, perhaps the first time the US backed away from a commercially sound idea.

While initial versions of new inventions are almost never perfect, over time the usual process of improvements would also have turned supersonic jetliners into planes as optimized and cost-efficient as our current jets but then several times faster. In fact, designs for a third generation of jetliners, so-called hypersonic aircrafts, were already made during the 1950s and 1960s, based on scramjets. Scramjets, which do not have compressor fans, scoop up oxygen as the plane travels through the atmosphere at speeds of

Mach 5 (5 times the speed of sound) and above. The high speed itself compresses the air within the nozzle before being diffused into the combustor. In this way, NASA's XX-43a scram jet design reached Mach 9.6 in 2004. Hypersonic passenger planes could fly to Australia from Northern Europe in less than five hours.

Even now we occasionally read about new, hypersonic passenger plane designs and how with such aircrafts we will soon fly from New York to Tokyo in a couple of hours[40]. There are artist impressions and discussions whether or not such planes should have windows, what their impact is on the environment, etc. But do we really believe this or is it just distractive reading in some waiting room?

Fly me to the moon

While cars and airplanes became very much the face of the 20th century, it is really the spaceship that captured the imagination of the 1960s as the ultimate travel experience. Now, in 2011, it is difficult to imagine how excited people were when in the summer of 1969 the first humans set foot on the moon. At the time, there could not be any doubt that this was merely the beginning of what was going to be a glorious path to the stars. Indeed, for Asimov in his 1972 *The Gods Themselves* a permanent moon base established early in the 21st century was a given and his description as to how this developed into a world of its own over the next half a century is as logical as it is entertaining[41]. And even as late as 1996 I was asked to participate in a NASA advisory panel to evaluate the necessary research and possible countermeasures regarding the radiation risk associated with a mission to Mars, at that time anticipated to occur between 2015 and 2020. And in our report we took that very seriously[42].

From a technological point of view, all this optimism is understandable. After all, if the United States could manage a moon landing a mere seven years after John Glenn became the first American to orbit the Earth, what couldn't we do over the next decades, especially when teaming up

40 Morris S. "The hypersonic plane designed to reach Australia in under five hours". *The Guardian*, February 5, 2008.

41 Asimov I, *The Gods Themselves*, Bantam 1972.

42 "Modeling Human Risk: Cell & Molecular Biology in Context", Report Number LBNL-40278, available to the public from the National Technical Information Service US Department of Commerce 5285 Port Royal Road, Springfield, VA 22161.

with other nations to share some of the expense? More than 4 decades after the first moon landing, that question is still very real and we do not really know the answer. Of course, what we do know is that the widely expected expedition to Mars never materialized and that the only major international space collaboration, the International Space Station is still being assembled in low earth orbit. Why did we never go to Mars and what could have been accomplished when instead of abandoning our bridge to the stars NASA would have vigorously pursued manned space programs of exploring our solar system and, some day, further outwards to neighboring stars?

The reason often mentioned is cost. It is expensive to send humans into space and to set up colonies on the moon and Mars. Obviously, there are lots of destinations for all that money, many more worthy than a manned Mars mission. As some would argue, it is more cost-efficient to send robots for space exploration. And it has to be said that the unmanned Pioneer and Voyager expeditions to the edge of our solar system and beyond have been very successful. However, there are at least two reasons why sending robots is less than satisfactory. First, robots have their own problems. They are far less capable as the human brain in responding to unforeseen problems and therefore in exploring unknown territory[43]. Second, and perhaps more importantly, in order to let humans travel in space in a meaningful way, we need to break technology barriers and that is exactly what the moon landing program did. There is a long list of inventions associated with the Apollo program, varying from recycling fluids and flame-resistant textiles to WD-40, which was developed for protecting rockets from corrosion while sitting on humid Florida launch pads.

The immediate reason for the collapse of space exploration is the never-fulfilled promise of cost-efficient, reusable launch vehicles. As we all know from the science fiction books and movies, the obvious way of traveling into space is by a rocket ship that will take off, do its thing in space, and then return to Earth for refueling and reuse. We cannot do that because rocket engines are not powerful enough to bring us into space in one piece. This is why rockets come in multiple stages, each of which with its own engines and propellant. The Space Shuttle, which made its maiden flight in 1981, was supposed to change all that. As the name suggests, the shuttle is reusable in the sense that its winged orbiter is capable of re-entering the earth's atmosphere, descent and land like a normal plane. It is capable of carrying

43 Kean S. "Making smarter, savvier robots". *Science* 329, 508-509, 2010.

large payloads with up to 8 astronauts into orbit. As such, it is critical to supply the International Space Station because the alternative, the Russian Soyuz spacecraft has a much lesser capacity. Unfortunately, two dramatic accidents, one in 1986 with the Challenger and the second in 2003 with the Columbia, costing the lives of all 14 astronauts, made it difficult to continue the program. Perhaps partly because of these accidents the costs of each launch became dramatically higher than originally anticipated. The program is now bound to be ended in 2011, leaving us without an economic way for bringing heavy payloads into space.

Apart from being the key to space exploration, launch vehicles are needed to put communication and weather satellites into space, which have a direct impact on our quality of life. The key problem is rocket technology, which has essentially remained the same since Sputnik. Indeed, over the last decades launch cost into low-Earth orbit has remained high, with the technologies used to design and build the current launch systems similar to those that were used thirty years ago. Current expendable launchers are based on the same approach as those developed over half a century ago by space pioneers like Konstantin Tsiolkovsky, Robert Goddard, and Wernher Von Braun. With the space shuttle gone, our options for exploring space seem dim.

Heavy-lift launch vehicles are regarded by many as the key technology for an aggressive, sustainable program of human spaceflight beyond low Earth orbit. In 2006, NASA announced plans to develop a new generation of space vehicles for transporting humans to low-Earth orbit and the International Space Station, and eventually to the moon and Mars. An international base camp was planned on the moon by 2024. Could it be that Asimov's expectations were not so unrealistic after all?

However, the fate of this new program, called *Constellation*, is currently in limbo due to budget constraints. Canceled outright by President Obama, parts of it have now been resurrected[44]. But even if it would eventually be continued the new, Orion orbiter and Ares launch vehicles to replace the space shuttle are basically a return to the days of the Saturn V launcher and Apollo orbiter. If manned human space exploration had not been abandoned, the development of economic launch capability would have been a

44 Mann A. "NASA human space-flight programme lost in translation". *Nature* 472, 16-17, 2011.

top priority. In its absence, progress lagged behind and this is not for lack of good ideas.

It is true that the process of developing new launch vehicles is expensive and technically challenging. But does this explain why the past decades have not seen major technological breakthroughs? After all, the Saturn V that brought us to the moon was developed in a mere 5 years and remains the largest and most powerful launch vehicle ever. Russia's *Energia* had about the same power and hypothetical future versions of this rocket would have been significantly more powerful than the Saturn V. However, its further development was canceled. Also the development of the Space Shuttle, the first orbital spacecraft designed for reusability, took only 5 years; the shuttle program was formally launched on January 5, 1972.

Although industry made all components of the rockets, a lack of industry initiative and risk taking is obvious. Like in the energy sector and the car or aircraft industry, in the absence of huge government programs very little is happening. More recently, a few entrepreneurs, e.g., Jeff Bezos, Elon Musk (the same who kick started the electric car industry), began investing their own money to build a new generation of low-cost and at least partly reusable rockets that can put payloads and people into space. Elon Musk's company, Space Exploration Technologies (SpaceX), succeeded in launching its unmanned Dragon capsule into orbit on top of its Falcon 9 rocket and recovering it from the ocean. We have of course seen all that in the 1960s with the Mercury, Gemini and Apollo orbiters, but this is the first private initiative to pursue a manned commercial space program. While one day this may lead to a new class of orbital transportation vehicles it is not likely to happen soon for the simple reason that the technological problems are enormous and their solution an expensive proposition.

It is obvious from our most recent history that a lunar base could already have been realized, exactly as predicted in the 1960s. The fact that it is still not there has nothing to do with physical limits to technological progress or a lack of human brain power, but because of reluctance to cough up the necessary funds. Interest in space adventures simply waned during the 1970s. This is in spite of some very articulate supporters. The most prominent among them was Carl Sagan (1934–1996), a highly successful popularizer and communicator of progress in science and technology. His 1994 book

Pale Blue Dot: A Vision of the Human Future in Space[45] is a cogent argument for expanded space travel and exploration.

An interesting model as to how human space travel could still flourish is the public–private partnership instigated in the 15th century by the Portuguese prince Henry the Navigator (1394–1460), which eventually established the Portuguese and Spanish colonial empires. Under the direction of this visionary the Portuguese first developed a new and much lighter ship, the caravel, which would allow sea captains to sail further, faster and much more efficiently. Columbus' three ships on his now famous first trip to the Americas were caravels. In a perfect example of public–private partnership, Henry sponsored the voyages down the coast of Africa that would eventually lead to the circumnavigation of the southern tip of the continent, now known as the 'Cape of Good Hope', and Vasco da Gama's 1498 trip to India. Also in those days, of course, there was great uncertainty whether it would all be worth the risks. Yet Henry's efforts transformed the world and created new sources of enormous profits. As of now, no one has serious doubt that a fully operational, efficient launch vehicle would revolutionize space transportation. Opening up the solar system to human use would be the biggest change since people learned how to crisscross the oceans, thereby establishing our global economy.

45 Sagan C. *Pale Blue Dot: A Vision of the Human Future in Space.* Ballantine, New York, 1994.

Chapter 4. Modern Medicine: Stuck in a Rut?

In 2003, I was invited to present some of my work in Cambridge, England, at a conference called Strategies for Engineered Negligible Senescence, or SENS for short. SENS is an initiative by the colorful and controversial Aubrey de Grey, as part of an attempt to show that aging can be cured[46]. His arguments are that scientific progress in medicine is so fast that someone who is born today will no longer die from natural causes and could essentially be immortal. Immortality, of course has been an obsession for humanity for ages, albeit almost exclusively for the well off. Powerful men (they were almost exclusively men) such as Gilgamesh, the ancient King of Uruk, and Qin Shi Huang, China's first emperor, did their utmost to find the elixir of youth, naturally in vain.

Aubrey's approach is different. For him aging is an engineering problem and simply by identifying all the causes of aging it should be possible to design remedies forestalling disease and pushing back death. And he does have a point. Aging, disease and death are processes amenable to scientific understanding and intervention. This is not based on myth, magic or charlatanism — bastions of those who purported to delay or 'cure' aging in olden times. Rather, it is driven by the impressive progress in medicine that has swept the world since the industrial revolution began at the end of the 18th century. Widespread application of new medical technology, from disinfection methods to vaccination and antibiotics virtually eliminated major

46 http://www.sens.org/

killers, such as smallpox, tuberculosis and polio, greatly reducing childhood death and disability, and permitting surgery on a scale never before seen. In the US, for example, life expectancy at birth increased from about 50 years in 1910 to its current level of almost 80 years. For Japanese females, life expectancy is now more than 85 years.

Similar to progress in information technology, innovation in medicine has been breathtaking and it seems far from unrealistic to assume that further, ever accelerating progress will indeed allow us to cure aging and attain immortality. Aubrey proposes that the therapies that will emerge will delay aging to such an extent that there is time to seek more effective therapies later on. He describes a point on this ladder of progress where medical advances will be so quick that death can be avoided altogether. Aubrey calls this moment 'Longevity Escape Velocity', which is very similar to the concept of singularity, when intelligent machines become our superiors and design ever more capable machines, leading to the intelligence explosion postulated by the already discussed futurist Ray Kurzweil. His name will come up again later in this book.

In an article in *Nature* my friend and collaborator Judy Campisi[47] and I argued, in 2008, that more research is needed on the causes of aging before we can evaluate if abrogation of human senescence is a realistic prospect. This careful conclusion is of course what you would expect from two serious scientists. Yet we also noted that there is no scientific reason for not striving to cure ageing, which implies that such a cure is in principle not impossible. Unfortunately, similar to the situation in the two areas of technology development we just discussed, medical technology appears to be characterized by a slowdown. In an article in *Perspectives in Biology and Medicine*, the Indian physician Indraneel Mittra calls this 'stuck in a rut'[48]. What happened?

The heroic age of medicine lasted from the 19th to the mid-20th century. It was dominated by independent, brilliant scientists, such as Edward Jenner (1749–1823), the pioneer of the small pox vaccine, William Morton (1819–1868), who developed surgical anesthesia, Louis Pasteur (1822–1895), who showed that microorganisms were the cause of human disease and developed the first rabies vaccine, Wilhelm Rontgen (1845–1923), who de-

47 Vijg J, Campisi J. "Puzzles, promises and a cure for ageing'. *Nature* 454, 1065-1071, 2008

48 Mittra I. 'Why is modern medicine stuck in a rut?" *Perspect Biol Med*. 52:500-517, 2009.

veloped X-ray diagnostics, Alexander Fleming (1881–1955), who discovered penicillin as the first of many future antibiotics, Sidney Farber (1903–1973), who developed cancer chemotherapy, Willem Kolff (1911–2008), the inventor of the kidney dialysis machine and the father of artificial organs, and Michael DeBakey (1908–2008), who pioneered cardiovascular surgery. Today, the situation is very different.

Medical research is now no longer dominated by talented physician-scientists developing treatments for their patients but by professional investigators who no longer see patients, often work in large centers, funded by billions of dollars from government, industry, universities and foundations. This gigantic research machinery still produces results, but much slower and at an incomparably greater cost-to-benefit ratio than was the case with the likes of Jenner. In 2007, Pfizer, the world's largest drug company, spent about $7 billion annually to develop new medicines. Its research operations encompassed eight major campuses with 14,000 scientists on three continents. However, from 2000 to 2005, Pfizer's laboratories created only few new drugs, including Caduet, a heart drug which basically combines two older medicines[49].

The situation is very similar for other giant drug companies, such as Merck and Schering-Plough, who developed Vytorin, a combination of a statin (see below) and a drug that blocks cholesterol absorption from the intestine. Abraxane, a newer version of the old cancer drug taxol, is made by Abraxis BioScience and costs 25 times as much ($4,200 a dose) as a generic version of the older medicine but does not help patients live longer (though it does shrink tumors in more patients) and has similar side effects[50]. The story is the same for Folotyn, an improved version of the decades old cancer chemotherapeutic drug methotrexate[51]. Marketed by Allos Therapeutics, it cost an astounding $30,000 a month only for a reduction in tumor size; there is no evidence that it actually extends life!

It was not always like that. Most current cancer drugs were developed between the 1950s and 1970s, based on the principle that blocking cell proliferation with agents that damage the genetic material preferentially kills

49 Berenson A. "A Pfizer Scientist Sees Research Dividends Ahead. *The New York Times*, July 18, 2006.

50 Berenson A. "Hope, at $4,200 a Dose". *The New York Times*, October 1, 2006.

51 Izbicka E et al. "Distinct mechanistic activity profile of pralatrexate in comparison to other antifolates in in vitro and in vivo models of human cancers". *Cancer Chemother Pharmacol.* 64, 993-999, 2009.

tumor cells with most non-dividing, normal cells spared. (These drugs do affect normal, dividing cells as well, such as the cells in hair follicles, digestive tract and bone marrow, which explains why patients undergoing these therapies usually experience hair loss, inflammation of the digestive tract lining and decreased blood cell production.) While there are great scientific advances in understanding the Achilles heels of tumor cells, affecting only their proliferation and survival but not those of normal cells, progress in getting these drugs to market is very slow.

We see a similar picture for treatments for other diseases. Angiotensin-converting enzyme (ACE) inhibitors, drugs for the treatment of high blood pressure, a major risk factor for heart disease, were rationally designed based on basic research in the 1970s[52]. Similarly, the development of statins, i.e., drugs that block the body's production of cholesterol thereby markedly reducing morbidity and mortality from vascular disease, including heart disease and stroke, was instigated by the discovery by Michael Brown and Joseph Goldstein of the LDL receptor and its central role in extracting cholesterol from the bloodstream[53]. Such statins as Pfizer's Lipitor became the best-selling pharmaceuticals in history, with many healthy people who happened to be diagnosed with elevated cholesterol (I am one of them) also taking these drugs.

Similar to the situation for cancer chemotherapeutic drugs, finding superior successor drugs for heart disease proved difficult. This is not because of a lack of ideas. For example, cholesteryl ester transfer protein (CETP) inhibitors, drugs that raise the level of HDL (the good cholesterol) are logical candidates. My friend and colleague Nir Barzilai, who studies human centenarians, those lucky individuals who live to be 100, found that lower CETP is likely to be a key factor in the long life span of these individuals[54]. This is not surprising since it affects such major human diseases as cardiovascular disease, diabetes and high blood pressure.

Pfizer was the first to develop a CETP inhibitor, called torcetrapib. Unfortunately, this drug did not survive its Phase III clinical trial[55]. In the

52 Vane J. The history of inhibitors of angiotensin converting enzyme. *J Physiol Pharmacol.* 50, 489-498, 1999.

53 Krieger M. The "best" of cholesterols, the "worst" of cholesterols: A tale of two receptors. *Proc. Natl. Acad. Sci USA*, 95, 4077–4080, 1998.

54 Atzmon G et al. Lipoprotein genotype and conserved pathway for exceptional longevity in humans. PLoS Biol. 4, e113, 2006.

55 Berenson A. "Pfizer Ends Studies on Drug for Heart Disease." *The New York Times*, December 3, 2006.

study, which involved over 15,000 patients with coronary heart disease (CHD) the torcetrapib group showed a 25% higher risk of CHD than the control group (taking a statin only). Also other drug companies have CETP inhibitors in clinical trials. However, Pfizer spent 800 million dollars developing torcetrapib, which made it the most costly failure in the industry. The problem is likely to involve other effects of this specific drug, rather than lower CETP levels. Nevertheless, Pfizer's failure raised the bar significantly for gaining approval of other drugs in this still highly promising line of potential blockbuster drugs. Furthermore, it will likely drive the pharmaceutical industry to stay away from such risky endeavors and, instead, focus their attention on making slight modifications in existing drugs, trying to extend patent protection and hang on to their branded drugs longer, for example, by making deals with generic drug makers. Meanwhile, the once gushing drug pipelines gradually run dry, turning innovative companies into rent-seekers, protecting their businesses behind multiple barriers to entry.

The failure of the medical research complex to develop new, truly innovative medicines is all the more disappointing in view of the enormous resources that went into new research over the last decades. This is best illustrated by cancer. In 1971, President Richard Nixon declared a war on cancer and since then more than $100 billion of taxpayers' money has been spent on combating the disease. However, cancer still takes the lives of more than half a million Americans each year with death rates only going down very slowly[56]. While there has been tremendous progress in understanding the nature and causes of cancer, a definite cure looks far away and some have given up hope that we may ever get there[57]. Why is this?

56 Leaf C. Why We're Losing The War On Cancer [And How To Win It]. *Fortune* 149, 76–97, 2004; Miklos GLG. The Human Cancer Genome Project—one more misstep in the war on cancer. *Nature Biotechnology* 23, 535 – 537, 2005.

57 Of note, in a recent assay in the journal *Cell*, with Daniel Haber as lead author (Haber DA, et al. "The evolving war on cancer". *Cell* 145, 19-24, 2011), it was argued that recent advances in blocking specific cancer drivers, so-called targeted therapy, are now beginning to bear fruit and may at last lead to victory. However, even these authors acknowledge that the necessary shift away from the classical formula for testing drug efficacy on groups of patients assumed to be similar to a more personal approach with a fragmentation of the market into cancer sub-types may be resisted by current business models and drug regulatory routines. There is also likely to be a lack of public support for the implementation of these innovative yet initially expensive new models for therapy. This is entirely in keeping with the thesis of this book.

Some would argue that currently incurable, complex diseases are not the equivalent of healing broken bones or getting rid of a simple infection. One possible explanation for the slow progress could therefore be the enormity of the problem, requiring much more research. The need for more research is indeed a common refrain in the world of biomedicine, but is this always justified? Progress in science has been immense and we now know infinitely more about what causes complex diseases than a few decades ago. Based on that knowledge we should be able to develop successful interventions, and to some extent, we have. Why are the returns on investments in medical research so disappointing? If currently incurable diseases like cancer would truly place our scientists for greater hurdles than equally life-threatening diseases from the past, such as tuberculosis, we would expect to see this reflected in a lack of creative solutions from our scientists. However, this is not the case. For preventing heart disease and stroke many new approaches come out of our laboratories. I already discussed CETP inhibitors for fighting heart disease and a major candidate to turn at least some of us into centenarians. But there are many good candidates for lots of diseases. For example, researchers have developed nanoparticles with a synthetic form of HDL to soak up the bad cholesterol that causes the formation of artery-clogging plaque[58]. This would be a great alternative for CETP inhibitors, just for in case these will really prove not to work.

There is a large variety of good ideas for curing human disease, some based on the manipulation of the patient's own defense systems, such as the immune response. An example close to my heart is the work of my wife, Claudia Gravekamp, also a professor at Einstein, who developed an ingenious strategy to cure cancer based on bacteria (made harmless) that infect tumor cells in the body and make them display proteins against which all of us have been vaccinated during childhood, e.g., proteins of viruses, such as rubella, measles and polio. Once vaccinated against these diseases during childhood, the immune system remembers them until very late in life in the form of so-called memory cells that are capable of mounting a quick and vigorous response the moment they see one of these proteins again, but now displayed by a tumor cell. This approach circumvents the main reason cancer is often so difficult to fight, i.e., the capacity of the tumor cell to hide

58 Thaxton CS, Daniel WL, Giljohann DA, Thomas AD, Mirkin CA. "Templated spherical high density lipoprotein nanoparticles". *J Am Chem Soc.* 131, 1384-1385, 2009.

itself from the immune system. Now exposed through these childhood vaccination proteins they become an easy target for the immune system. While I of course tend to think my wife is unusually smart, there are simply too many other examples of new, potentially successful approaches for curing disease to seriously consider a lack of human ingenuity as the reason for the disappointingly slow progress. For that, we need to look elsewhere.

The first time I realized that life was not as simple as developing a drug, test it and begin treating patients was in late summer of 2008, shortly after I had become the new Chairman of the Department of Genetics at the Albert Einstein College of Medicine in New York. At that time Ganjam Kalpana, a professor in my new department, came into my office to seek advice about how to speed up clinical trials for a new treatment she had developed for rhabdoid tumors, a rare and highly malignant childhood cancer. Trained as a basic scientist, I have found that an important part of my job description at Einstein is to greatly accelerate what we call translational genetics, to bring new discoveries in human disease genetics into the clinic as fast as possible. Kalpana's story was sobering. She had discovered that rhabdoid tumors in mice could be effectively cured by inactivating a specific molecule, called cyclin D1[59]. Gone out of control, it is this molecule that drives the growth of cancer in these children. Luckily, Kalpana found that at least two existing drugs (Fenretinide and Flavopiridol), which were already in clinical trials for treatments of other cancers, effectively suppressed cyclin D1.

Since Kalpana's findings had been published in the scientific literature, she began to receive lots of email messages from desperate parents of children with this disease. Treatment with one of these two drugs seemed straightforward. After all, they had already undergone successful clinical trials to evaluate their safety, determine safe doses, identify side effects and look at their effectiveness. So, Kalpana expected clinical trials for this particular cancer to begin very quickly. That was 2007! Since then very little has happened. One of the main reasons is a lack of interest from the drug industry. Developing new treatments for cancer is a costly business, not so much because of the research that needs to be done but mainly because of the regulatory burden. In spite of Kalpana's clear-cut results, they simply did not see a profit in developing these drugs for such a rare disease.

59 Alarcon-Vargas D et al. "Targeting cyclin D1, a downstream effector of INI1/hSNF5, in rhabdoid tumors". *Oncogene* 25, 722-734, 2006.

Unfortunately, examples of abandoned medical discoveries because of a lack of profit abound[60] and due to the ever-increasing regulatory burden help from alternative sources of financial support are a long shot[61]. Because it is the drug industry that is primarily responsible for giving us new treatments for disease, Kalpana was now confronted with a situation where she had to do the work of the drug developers. Since she is not trained to do that and does not have the resources, the patients need to wait and there is a fair chance that not a single patient will ever get treated with Kalpana's new candidate drugs for rhabdoid tumors. The lack of follow-up by the pharmaceutical industry of so many basic science discoveries prompted the director of the National Institutes of Health (NIH), Francis Collins, to create a center for drug development, the *National Center for Advancing Translational Sciences*. According to the NIH web site the purpose of the new center is not so much to take over a job that the drug industry is not doing to everybody's satisfaction, but to 'de-risk' drug and therapeutic development projects thereby rendering them more attractive for private sector investment[62].

Another major reason for the disappointing results in developing new, successful interventions to cure current killer diseases is the increasingly defective connection between basic science and the clinic. In the old days, doctors themselves were not hesitant to take initiative in developing new treatments. For example, the aforementioned physician Willem Kolff invented the artificial kidney in the early 1940s, which evolved into modern dialysis machines for cleansing the blood of people whose kidneys have failed. His first artificial kidney, made by using sausage casings and orange juice cans, caused the death of the first 15 people placed on the machine. However, he made improvements, including the optimum use of blood thinners to prevent coagulation[63].

Nowadays, doctors are too busy with the day-to-day activities of seeing patients and the intricacies of the reimbursement process to have much time for initiating their own research. Those who manage to still get in-

60 Begley S. "From bench to bedside: Academia slows the search for cures". *Newsweek*, June 15, 2009.

61 Morice AH. "The death of academic clinical trials". *The Lancet* 361, 1568, 2003; Akst J. "EU trial rules stall research". *The Scientist*, November 17, 2009; Stewart DJ et al. "Equipoise Lost: Ethics, Costs, and the Regulation of Cancer Clinical Research". *J. Clinical Oncology* 28, 2925-2935, 2010.

62 http://feedback.nih.gov/index.php/faq-ncats/

63 Henderson LW. "Obituary: A Tribute to Willem Johan Kolff, M.D., 1912–2009". *J Am Soc Nephrol* 20, 923–924, 2009.

volved in research prefer to help unraveling the complex molecular pathways that play a role in disease processes. Indeed, even the doctors are now focused on basic research and not so interested in the more practical aspects of curing patients. In his recent essay *Why is modern medicine stuck in a rut?*, Indraneel Mittra calls this the 'hegemony of molecular science'. He wonders 'which of the countless pieces of molecular information has helped to make his patients better'.

Being a molecular geneticist myself, I sometimes also wonder why so much time and effort is being spent, often by MD's or MD/PhDs before they go into clinical practice, on adding yet another gene or protein molecule to the already endless list of names defining our molecular heritage. Yes, this is critically important and we need to spend time on it, but surely not to the extent that doctors themselves give preference to working on these jigsaw puzzles rather than using their own power of judgment to make their patients better. Indeed, we already know most of the genes that are important in human disease. For example, most or all the molecular pathways that guide cancer-inducing signals through cells are known. It is now time to apply this knowledge through integrated programs in which both basic scientists and physicians participate. Meanwhile, it would be important for doctors not to overestimate the power of molecular knowledge in helping to cure their patients. Their job should be to use molecular information in a pragmatic way, point out critical unknowns to their colleagues in basic science and develop practical interventions. Their job is not to increase the amount of molecular information for information's sake.

Unfortunately, the unraveling of molecular pathways, not their use in clinical interventions, is the way to get scientific recognition, publications and promotion. Ironically, to publish a paper in a top journal describing merely a new cure for cancer has much less priority than hard core science providing an interesting mechanism. This is also true for the NIH study sections where your grant proposal is reviewed. Hypothesis-driven research proposals are valued much more highly than something mundane as curing a major disease. Through the years, medicine has become stratified and divided itself into a research and a clinical branch, each highly professionalized and focused on itself.

Albeit important, an eagerness of doctors to emulate PhDs and fill in the gaps in our molecular universe rather than developing practical applications of all that knowledge and cure their patients, is not the only reason for them to abandon the more practical approaches of yore. While everybody is still highly impressed by ancient physicians like Kolff and DeBakey, as well as even earlier generations as Pasteur and Fleming, it goes without saying that nowadays an artificial kidney could never, ever be developed. Indeed, already after the first fatality the approach would have been abrogated and unlikely to ever have been continued.

Progressively tighter regulations on new drugs are providing an incentive to be conservative. Sometimes drugs are withdrawn from the market because of one or two serious adverse events out of tens of thousands of patients. This happened with the arthritis pain killer, Vioxx. Vioxx is a so-called nonsteroidal anti-inflammatory drug (NSAID), the same as Ibuprofen, but less likely to cause bleeding because it specifically inhibits cyclo-oxygenase 2 (COX-2) rather than both COX-1 and 2[64]. Merck withdrew Vioxx from the market in 2004 because a new study had found a higher rate of heart attacks and strokes in patients taking the drug than in those on a placebo. However, the vast majority of arthritis patients benefit from this kind of drug, all of which have been linked to cardiovascular risks[65].

This is not to say that regulation is a bad thing. Avoiding unwanted effects of new treatments is obviously of critical importance. Unfortunately, adverse effects, even fatalities, among subjects of clinical trials cannot be entirely prevented and the occasional withdrawal of a drug for these reasons is part of the process of innovation. Tolerance for such mishaps has significantly decreased lately and the consequence is that highly promising treatments find themselves subject to much longer development times or even cancellation at the slightest sign of trouble. Increased animal research before clinical trials could help minimizing fatalities, but this also is heavily regulated and not popular with the public. All this greatly increases costs, drive non-sponsored clinical trials out of practice and increase the attractiveness for the drug industry to retrench and limit themselves to slight modifications of existing drugs.

Intolerance for even a minimum of casualties, unavoidable in testing powerful new therapeutic approaches, has now reached a level that effectively constrains any further attempts to seriously innovate in this area.

64 COX-1 is necessary for homeostasis of the gastrointestinal mucosa. Inhibition of COX-2, which is activated during inflammation, does not affect the gastrointestinal mucosa.

65 Gallagher RM. "Balancing Risks and Benefits in Pain Medicine: Wither Vioxx". *Pain Medicine* 5, 329-330, 2004.

This was made clear to me when listening to one of my colleagues in the Department of Genetics at Einstein, Jack Lenz. Lenz is a professor in genetics and works on gene therapy. The story he told me was about several clinical trials for testing the replacement of the defective gene in patients with different genetic diseases. Some of these patients developed cancer as a consequence of the treatment. This has to do with the fact that it is still very difficult to target these therapeutic genes so accurately that they will always end up in a place where they cannot influence other genes. It turned out that in some cases the healthy genes landed near a gene that if activated could promote cancer[66]. And, indeed, some patients did get cancer!

The result of these tragic events was that for the time being such clinical trials were halted and gene therapy lost its promise in the eyes of many. However, while this suggests that gene therapy as a new and promising technology for curing heritable diseases is an abject failure, nothing could be further from the truth. In fact, 75% of all patients did not suffer adverse reactions to the therapy and were cured from their disease. Most importantly, and this is why Jack and I had this conversation, he and others are working hard to re-design the reagents used in this therapy to make them more specific and safe[67]. However, they are now confronted with a public and a regulatory body that lost their nerve and make it much more difficult to further continue this work. Unlike Kolff, who could improve his experimental dialysis machine, current medical researchers are often denied attempts to make new therapies come to fruition because the regulatory authorities decide that the risks of fatalities (which cannot be reduced to zero) outweigh the fact that the far majority of the patients participating in the trials were actually cured of their devastating disease.

Stem cells, another promising new avenue of treatment will undoubtedly suffer many of the same problems, with even slight setbacks not tolerated. But in this case there are other problems as well. Stem cells, which can differentiate in essentially all cell types in the human body, have great potential in medicine as replacement cells for a variety of degenerative diseases[68]. Stem cell-based, regenerative therapies have been proposed for dis-

66 Williams DA, Baum C. "Gene therapy – new challenges ahead". *Science* 302, 400-401, 2003.

67 Naldini L. "Inserting optimism into gene therapy". *Nature Medicine* 12, 386-388. 2006.

68 Schwartz RS. "The Politics and Promise of Stem-Cell Research". *The New England Journal of Medicine* 355, 1189-1191, 2006.

eases such as Parkinson's disease and Alzheimer's disease. However, stem cells can only be obtained from human embryos and some are of the opinion that to use an embryo for that purpose is ethically wrong. Of course, lots of early-stage human embryos are discarded in clinics for in vitro fertilization and 70% of Americans favor the further development of stem cell therapies. Nevertheless, in 2001 the administration of President George W. Bush implemented strict limits on the use of human embryonic stem cells, and even under his successor, Barack Obama, who is generally favorable to stem cell research, the U.S. District Court for the District of Columbia ruled in 2010 to halt federally funded embryonic stem cell research[69]. Because of the uncertainty this generates scientists seek other areas of interest, students pursue other career paths and investors spend their money elsewhere. Imagine what would happen if antibiotics still needed to be invented! Many of us would still die of infection.

On the face of it, increased regulation reflects how careful we have all become in protecting each other from mishap. However, there is a trade-off. Increased regulation suppresses innovation and could eventually diminish healthy competition in drug development with only few large companies making more of the same remaining. Most of the costs involved in bringing a new drug to patients come not from the initial discovery research but from clinical testing and regulatory submission. This has now become so exhaustive that even doctors who do want to spend their own time on testing new drugs are no longer able to do that. In the past things were relatively simple, but nowadays extensive protocols need to be prepared for the institutional review board and then every time a patient in the study has an adverse reaction (often without any relation to the drug or procedure tested), the review board needs to be notified. Most doctors are unable to do all that. Hence, a major reason for the disappointing performance of our biomedical establishment in developing new, breakthrough therapies is the cost of regulation[70].

A consequence of increased regulation is the slowdown in new drugs that hit the market. There has been a steady decline in the number of new active substances, also known as new chemical entities[71]. And, as expected,

69 Kaiser J, Vogel G. "Controversial Ruling Throws U.S. Research Into a Tailspin". *Science* 329, 1132-1133, 2010.

70 Akst J. "EU trial rules stall research". *The Scientist*, November 17, 2009.

71 Schmid EF, Smith DA. "Is declining innovation in the pharmaceutical industry a myth?" *Drug Discovery Today* 10, 1031-1039, 2005.

the number of new drugs approved by the Food and Drug Administration is slowing down (see Figure 4.1, on page 76)[72]. It is simply too expensive to produce a novel therapy, which is always a risky proposition. After all, while the harvest of new drugs lately is dismal, spending by the drug companies on research and development has almost quadrupled over the last decade. It has now become very difficult to bring the many basic discoveries for cures to fruition by translating them into therapeutics.

A third reason for a lack of breakthrough inventions in biomedicine is a lack of patients for participation in clinical trials. In the US, less than 10% of eligible patients actually enter clinical trials[73]. This is partly due to the very success of medical research. In contrast to the old days today there are therapies for about everything. For example, if in the past the diagnosis of cancer was often an immediate death sentence, nowadays death can often be postponed a long time and the patient has a choice. Hence, they prefer to stick with the old, proven therapies and are highly reluctant to participate in something new. Only terminally ill people see some benefit.

The desire to make us live longer and, most importantly, live healthier has been supported by some enormous collaborative research endeavors of which researchers in the past could only have dreamt. We now know all our genes, can visualize their activity in the different parts of the brain or in other organs, have tools to detect the slightest deviations from the norm, are capable of replacing someone's defective gene by a healthy copy, and many more. However, we do not seem to be able to reap the fruits of all this progress and this is not due to a decline in creativity. As we have seen for the energy and transportation areas of technology development, the true reason is the increasing difficulty to see new technology practically realized. In medicine, research and clinical practice are now worlds unto themselves allowing only occasional foraging in each other's territories. The crushing load of regulations, often disproportionate to the actual risks, and an already highly sophisticated system for preventing and treating disease, which creates reluctance of patients to expose themselves to the unavoidable risks that are always associated with testing new, potentially revolutionary treatments, make it difficult to develop and implement biomedical advances in clinical practice.

72 Corbett Dooren J. "Drug Approvals Slipped in 2010". *The Wall Street Journal*, December 31, 2010.

73 Young RC. "Cancer Clinical Trials — A Chronic but Curable Crisis". *The New England Journal of Medicine* 363, 306-309, 2010.

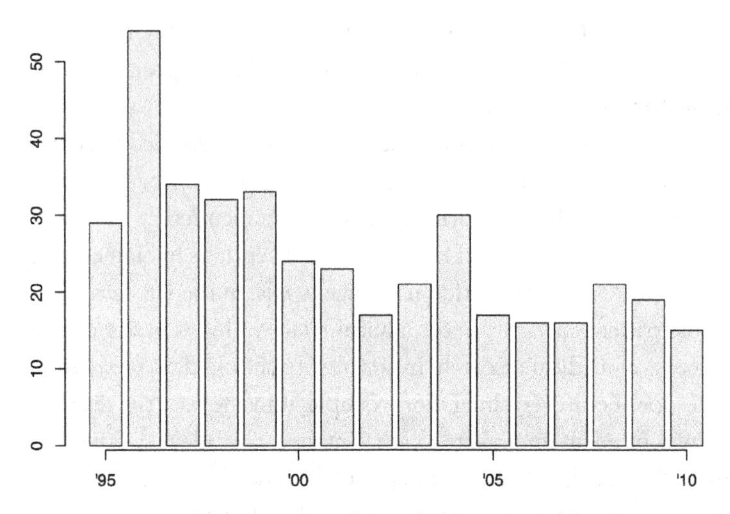

Fig. 4.1. Declining number of new drugs approved by the US Food and Drug Administration from 1995 to 2010. Source: FDA.

Chapter 5. The Information Revolution: Hope or Hype?

Anybody you would ask which area of technology has seen the greatest progress recently would likely refer to information technology (IT). While Charles Babbage might have been the first person to introduce the idea of programmable machines (in 1822), the first electronic digital computer was built by John V. Atanasoff and Clifford Berry at Iowa State University in 1942. Called the Atanasoff–Berry Computer (ABC) it was based on vacuum tubes. The vacuum tube, itself a macro-invention (1906), was essential for the early days of both the information and telecommunication industry. Its invention enabled the expansion and commercialization of radio, television, radar, sound reproduction, telephone networks and computing. Vacuum tubes were later replaced by transistors (1947), which developed into integrated circuits (1959). Integrated circuits enormously increased the number of components that could be built into a computer. The later Intel 4004 chip took this one step further by placing all the parts of a computer, i.e., central processing unit, memory, input and output controls, on one small chip: a microprocessor (1971).

The miniaturization of digital, programmable computing led to the first microcomputer (1972) and the subsequent personalizing and mass retailing of the instrument since the late 1970s (Apple and IBM), causing a wave of innovations in such major activities as word processing, bookkeeping and data storage. The subsequent invention of the internet and the World Wide Web brokered the match of the computer with the telecommunication industry, providing access to all kinds of information in seconds and minutes rather than a visit to a library or a place to consult records.

As we will see later in this book, the first real improvement in communication since the great empires with their excellent road systems and post-horses more than 3,000 years ago was the invention of the telegraph in 1837. While the telegraph has become obsolete, its successors, telephone (1876), radio (1895), television (1927) and the internet (1972) are now all converging into one single medium, which can be operated from the living room or office, the car, train or airplane, from a desktop or laptop computer or a handheld device. It is the computer that has made this convergence possible. The internet shrewdly uses existing telephone cables to create a global network of computers. Fiber optics networks greatly facilitated extensive use of this system, which is expanded even more by the current explosion of high-speed, 'broadband', wireless connections.

Telephone was originally firmly anchored to a home or office with abundant public phones on most street corners. The invention of the cellular phone in 1973 changed all that. Using radio, the mobile phone provided wireless access not only to other phones, but now also to computers through the internet. While initially slow, faster systems are now emerging allowing downloading everything from the internet onto a mobile hand-held device. The phenomenon of convergence leads to the mingling of everything technically imaginable into a single medium, whether TV, desk-top computer or portable multimedia player. These developments impact broadly on the way we live, work and entertain ourselves. For example, the convergence of telephones and global positioning systems (navigation systems that receive their signals from satellites or cell phone towers) allows you to get up-to-the-minute information on traffic, restaurants and gas stations. Some big cities now use webcams (digital cameras attached to a computer) to monitor and charge cars that enter the downtown area.

As I mentioned at the beginning of Chapter 1, in the eyes of many technology is now the same as information technology. This view, that IT has now transcended all human innovation, is not limited to the retail industry and the general public. For example, *The invisible future: the seamless integration of technology in everyday life*, edited by Peter Denning, gives us a singular view on what drives future technological development in the next decades[74]. From space travel to fishing and from nanotechnology to health care, the

74 Denning PJ (editor). *The Invisible Future*. McGraw-Hill, New York, 2002.

role of IT in our future is all encompassing. For me the question is not if this is true or not true, because there can be no doubt about the major role of IT in all these areas and more. The question is if IT will really make a mark on humanity that is much more than that left by the macro-inventions of the past, those that generated food security, world-wide transportation, abundant, fossil fuel-based mechanical energy. Is our future being shaped by IT to such an overwhelming extent that the disappointing progress in other areas of technology development is no longer important?

There can be absolutely no doubt that progress in wiring the world, interconnecting the different information carriers, and the great extension of computer power have significantly expanded our capacity of doing the things we were doing. For example, in the music industry, first radio, then gramophone and television and now the internet with Apple's iPOD and other portable music players dramatically increased everyone's opportunity to see and hear the artists perform. The same is true for the great sports events; it has never been easier to enjoy the World Cup soccer or the Super Bowl. This is of course reflected in the salaries of the top performers. The big stars of today, whether in sports, film or music, make a lot more money in real dollars than their predecessors in the past.

Similarly, in the financial sector IT has made an enormous difference. Stock trading is now mostly conducted by computers at high speed. Anyone can now buy stocks from anywhere in seconds with the simple click of a mouse. Based on the original invention of optical character recognition (1950), computers can speed-read news reports and then interpret and take the decisions to do the actual trading. The latter requires artificial neural networks the first practical version of which was invented in 1958.

Aided by deregulation in the 1980s the greatly increased computer power allowed the creation of novel instruments for earning money through speculation. Most of us would now consider much of this as scandalous self-enrichment and a danger to society. Yet, while derivatives are no macro-inventions, they are definitely innovations that follow logically from macro-inventions in the IT and telecommunication industry.

Apart from these two major areas of application, entertainment and financial services, IT is everywhere. We can pay our bills on line rather than by mail, shop electronically rather than by mail order, email rather than communicate by mail and meet friends on Facebook or Twitter. IT is now critically important in oil and gas exploration and the sophisticated computer programs in the airline industry are responsible for the unpleasant fact that now almost all planes are filled to capacity.

Progress in IT and telecommunication truly revolutionized society. Its dominance in technology development in the 20th century is underscored by the 24 macro-inventions in this area since 1930, which is almost 25% of the total. One macro-invention led to others, each of them spinning off a multitude of micro-inventions. However, there is reason to doubt that progress in IT is exponential as most futurologists assume. In fact, while the period from the 1950s through the 1970s saw 12 IT inventions, over the last 3 decades there were only five, i.e., file sharing, automatic speech recognizers, the first search engine, the World-Wide Web and instant messaging. Therefore, it is possible that like in other areas of technology development also the innovative IT era is gradually coming to an end. There are still lots of micro-inventions but no longer major breakthroughs. For example, take Apple's iPhone. Most people would consider that a major invention, stemming from 2007. In reality, all the ideas on which this sleek instrument is based were in place long before Steve Jobs came along realizing what consumers want. While truly a commercial and design masterpiece the iPhone is not a macro-invention. Most futurologists, however, look at it differently and are still convinced that skeptics like me will soon see the light. The reason for their optimism is progress in artificial intelligence.

Artificial intelligence

In his book *The Singularity is near*, the futurist Raymond Kurzweil argues that we are in the midst of a long-term pattern of accelerating change[75]. If some of the machines we make could even slightly surpass human intellect, they could improve their own designs in ways unforeseen by their designers, which would lead to a sudden, dramatic acceleration of intellect. It is that moment that has been termed 'the singularity', which was scheduled by Kurzweil to take place in 2045, when artificial intelligence would surpass humans as the smartest life forms on Earth.

75 Kurzweil R. *The Singularity Is Near: When Humans Transcend Biology*. Viking, New York, 2005.

While Kurzweil's predictions are unlikely to come true, I believe that he is correct by measuring real progress in IT by the advances we make in mimicking human-level thought in computers. There is little doubt that computers have superior decision-making skills as compared to humans. For example, in a comparison between humans and computer models as purchasing managers at more than 300 organizations, given the task of placing orders for computer equipment and software, the computer models performed significantly better in categories like timeliness of delivery, adherence to budget, accuracy of specifications and compatibility with existing systems[76].

The humiliating defeat of the world's best chess player, Garry Kasparov, in May 1997, by a computer program called Deep Blue in a highly publicized six-game match further underscores the edge computers hold over humans. It was the first time a computer had ever defeated a chess world champion.

With quantifiable data from past experiences, a computer algorithm will now easily outperform a professional's decision making skills. In general, computer models make more accurate predictions than humans do, which is due to their consistency and their much greater depth of data processing. In this respect, many putative managerial qualities like experience and intuition may be largely illusory. However, as soon as the underlying conditions change and the assumptions that went into the creation of those computer models are violated, reality catches up and smooth performance collapses. Unlike computer models, humans have the ability to recognize when something is wrong. It is this capacity that can probably best be described as 'intelligence', the ability to structure incoming information in a way that allows real understanding and a rational rather than predetermined response. In Isaac Asimov's 1950 collection of short stories, *I, Robot*[77] it is this human attribute that was eventually created in artificial form. In the novel this was called a 'positronic' brain, invented in 1982 (not far enough in the future, as is now apparent). A similar brain was introduced to us in the 1968 movie *2001: A Space Odyssey* by Arthur C. Clarke and Stanley Kubrick in the form of HAL 9000, an artificially intelligent computer, which maintains the space ship and is easily one of the most evil characters in film history.

76 Snijders C, Tazelaar F, Batenburg R. "Electronic decision support for procurement management. Evidence on whether computers can make better procurement decisions". *Journal of Purchasing and Supply Management* 9, 191-198, 2003.

77 Asimov I. *I, Robot*, Doubleday, New York, 1950.

Artificial intelligence has its origins in the famous 'Turing test' in 1950, when the British mathematician Alan Turing (1912–1954) conducted tests in which a person facing two computer terminals had to judge on the basis of questions and answers which of the two had a human behind it. If the person could not identify the terminal with the human, the machine was considered intelligent[78]. Regardless of whether this test is sufficient to judge if a machine is conscious and can think, all attempts to build systems replicating human mental capabilities that provide for speech, hearing, manual tasks and reasoning have thus far been unsuccessful, as anybody who ever tried to reason with the automated voice-dialogue system of a bank or airline can testify.

During the 1960s and 1970s, researchers began designing computer programs called 'expert systems'. Such programs were essentially databases with a set of logical rules. The results were disappointing and it is now realized that the difficulties in building conscious machines are more serious than ever imagined, the key single issue being a gross lack of knowledge of the structure and function of the human brain. In addition, the original artificial intelligence workers were handicapped by their lack of computing power. Since their invention in 1959, integrated circuits have doubled the number of components they can contain every two years. Already referred to in the Prologue, this is called 'Moore's law', after Intel co-founder Gordon E. Moore who first reported it in a 1965 paper[79], and has been responsible for a dramatically increased computer processing power. Progress in four separate fields, mathematics, neuroscience, computer science and psychology, are now contributing to a new synthesis that may finally provide us with the tools to break the barrier on the road to the thinking machine. This has already led to new theories on how the human mind works.

The human brain is the most complex structure known, and to mimic it, even to a small extent, is a daunting task. Using an IBM supercomputer, the activity of a small part of the brain, consisting of 10,000 nerve cells or neurons, is now simulated in the so-called Blue Brain Project[80]. The idea is to create a digital description of nerve cell connections. The fact that there may be 100 billion neurons in a human brain, each with as many as 10,000 connections explains why this first attempt involves only a tiny slice with

78 Turing AM. "Computing machinery and intelligence". *Mind* 59, 433-460, 1950.

79 Moore GE. "Cramming more components onto integrated circuits". *Electronics* 38, 1965.

80 http://bluebrain.epfl.ch/

a mere 10,000 cells. Nevertheless, the hope is that this may eventually lead to the creation of an artificial brain that may reveal how certain intelligent activities take place. It should be noted that this project is entirely based on traditional computation and programming techniques, which may not be applicable to the way we think. Still, there is progress also from other directions as well.

After Deep Blue had beaten the world's best chess player, IBM has now come up with a program that comes close to an artificial intelligence system, at least superficially. In 2009 they produced 'Watson', a supercomputer that appears to understand a question and can respond with a precise answer[81]. Watson is a question-answering system that can search through mountains of documents or enormous databases of information. It proved its mettle by winning, hands-down, a Jeopardy contest against experienced human contestants. It was able to take them on based on multiple algorithms designed to associate subjects with the words it finds most frequently associated with them. Such statistical computation is now possible because of the greatly increased computational power and the amount of online text. Indeed, similar systems could scour the World-Wide Web for interconnected information in response to a query. In essence, this would lead to a semi-intelligent search engine.

Applications of semi-intelligent software abound and vary from market research to financial planning and educational consulting. I could for example have used it for this book to do an automated fact check and make sure that I do not include material without proper reference. The idea is to have a system that spares you the cumbersome task to go through enormous amounts of information and, instead, selects some carefully weighed combinations that fit your purpose. However, it is important to realize that all these developments are still based on existing IT and have thus far not yielded systems that can reason in human fashion.

Innovations based on developments in artificial intelligence are visible in several areas. Robot cars, equipped with a wide variety of sensors and custom-written software, including machine-learning algorithms, can now drive automatically for hundreds of miles through rugged terrain. Ultimately, such artificially intelligent cars could lead to the development of driver assistance systems that keep civilian drivers, passengers and pedestrians

81 Thompson C. "What's IBM's Watson?" *The New York Times Magazine*, June 20, 2010

safe. Voice control systems are already found in some automobiles and advanced artificial reasoning techniques are routinely used in inexpensive video games to make the characters' actions more life-like. However, this is still far from the routine implementation of robots for carrying out a variety of specific tasks, including household chores and aids for the elderly, instead of specialized jobs, such as a dishwasher or a dryer. We have no robots that can clean up after dinner, load the dishwasher or replace a light bulb.

Nevertheless, while the current crop of humanoid robots may not exactly conform to our ideas of robots as super humans, we have come a long way. Enormous progress has been made in robot thinking, reasoning, creating, even improvising, as becomes obvious from a visit to the web site of the Humanoid Robotics Group of the Massachusetts Institute of Technology's Artificial Intelligence Laboratory[82]. Bill Gates, the famous ex-CEO of Microsoft, has said that the personal robot is about as far as the personal computer in the 1970s. This may well be correct, but there is an important difference. The personal computer could be easily integrated in our way of life while the personal robot will change our life in ways as yet difficult to foresee.

The first concepts of an *in silico* personal assistant were based on early work from the 1950s by Oliver Selfridge, another artificial intelligence pioneer, who died in 2008[83]. And it is true that to some extent current cell phones can be used as personal assistants. They can keep your calendar and email account, allow you to plot your journey from door to door, including public transport options, the latest traffic news and rerouting suggestions, find out the weather and manage accounts. However, thus far, a simple software program that can monitor its environment and make appropriate responses when unexpected changes occur has eluded us. The best that current systems can do for you is automatic re-scheduling of your appointments when your flight is late and informing your staff or family members.

In his 1995 book *Getting Digital*, Nicholas Negroponte, the founder of the Massachusetts Institute of Technology's famous Media Lab, writes about a future, soon to arrive, when we will all have digital 'butlers' who do things for us, like providing us with the type of news we are interested in, guiding us to restaurants where we like to eat, and screening through massive

82 http://www.ai.mit.edu/projects/humanoid-robotics-group/; Marantz Henig R. "The real transformers". *The New York Times Magazine*, July 29, 2007.

83 Licklider JCR, Taylor RW. "The Computer as a Communication Device". *Science and Technology* April, 1968.

documents to filter out what we are not interested in and summarizing the rest[84]. He also describes new, high-quality videoconferencing systems and novel ways to communicate with your computer in an effective way. Now, as of this writing, we still don't have electronic alter egos, do not like videoconferencing and are stuck with impossible operating systems for our computers. The latter is a good illustration of how a sophisticated business model can constrain real progress. Both Apple and Microsoft tend to keep upgrading their operating systems without bringing them any closer to Negroponte's smart computer. On the contrary, Windows Vista is a huge incoherent mess and only by carefully avoiding virtually all of its entirely unnecessary 'upgrades' do I manage to produce this manuscript.

Another example is internet TV. Cable TV service suffers from periodic service glitches and is expensive because it is only available in big bundles of channels, most of which you do not want. Instead, it would be ideal to program yourself from online sources, similar to what many of us now do with music. However, selecting your own TV shows from the internet is surprisingly difficult. To some extent, this is due to the distributors and producers who managed, better than their colleagues in the music industry, to protect their business models, to prevent shows, such as American Idol, and sporting events to be seen online. Most people still pay for traditional TV service, however, because it is simply too much of a hassle, especially for older people, to put together the package of shows and events they want to watch and actually can watch on the internet. For this reason there is as of this writing no sign that viewers leave cable TV in droves.

However, while Kurzweil's singularity may be very far away, progress in artificial intelligence is genuine and the practical optimism of Negroponte is entirely realistic. Unfortunately, it is questionable if serious breakthroughs in artificial intelligence will ever be a topic of major innovation and result in real products that will be made widely available. There is resistance from the public in adopting the changes that are inherent to genuinely new and revolutionary information technology, but also reluctance of IT companies to invest in truly novel and transformative concepts. Current business models greatly favor making electronic gadgets rather than investing in real breakthroughs for the simple reason that it is not a priori clear that the public is prepared to talk to a computer and trust its judgments in many, often personal, situations. We may not be ready to accept an electronic assistant

84 Negroponte N. *Being Digital*. Vintage, New York, 1995.

who is with us all the time and prefer computers to remain simple addition machines rather than walking and talking companions. With so many people not even prepared to accept automatic transmission in their cars, the barrier to letting computers infringe on our personal lives is formidable.

From a business perspective, therefore, a preference to continue investing in what appears to be working well rather than spending money on risky forays into unknown territories makes a lot of sense. Indeed, it is this smart business strategy, not risky innovation, which turned Apple into a corporate giant. In hindsight, many of the investment decisions of the past, from AT&T's transistor and mobile phone to Xerox's graphical user interface and ethernet made little sense and current business dictates would no longer allow the eminent scientists who worked in these labs the same free reign that led them to make all these discoveries. It would be naïve to see only the successes made possible in these anarchic times. There were many more failures, which is exactly why a business model not based on cutting-edge innovation is generally more profitable. Fortunately for us, it took a while before that sank in!

Chapter 6. A Crisis of Success: The Causes of a Technology Slowdown

Life in the 18th century for most people was not much better than that of their hunter-gatherer ancestors in the Stone Age and there was certainly a lot to invent that made economic sense. The average citizen in those days would have been astonished witnessing life in 1910, with its great steamships, dense networks of railroads and instant communication through the telegraph. Likewise, someone from 1910 would marvel at the stunning sight of the Los Angeles freeways, or the large jetliners taking off and landing at LAX in the 1960s. However, only very few teenagers of 50 years ago could have predicted that our current society would look so much like theirs. They would be terribly disappointed in seeing virtually the same airplanes, trains, motorways and hospitals as in their own time.

The status of progress in the four areas of technology development discussed in the previous chapters leaves us with the strong impression of a general slowdown, albeit not because of a lack of imagination or even new inventions. Instead, the problem appears to be further downstream and hinges on a number of constraints in implementing newly developed technology in society. This is obvious, even from simply looking around. Our cities have not undergone significant changes since the introduction of the skyscraper. Made possible by the invention of the steel frame at the end of the 19th century, the elevator (1853) and the tuned mass damper — a heavy ball suspended by cables inside the building's top to steady the building when it starts to sway — from 1909, those first skyscrapers were certainly a significant deviation from traditional building. The current rush to build ever taller skyscrapers, mostly in Asia, merely gives rise to faster, better elevators. They are no longer associated with differences in working or living as was clearly the case when they were first invented.

Fig. 6.1. Paolo Soleri's Babel arcology concept: a central core with apartments on a base of commercial and civil spaces. It should be able to house over half a million people in a structure that is 1,900 meters high and 3,000 meters wide. Courtesy Jeff Lewis.

However, it is not difficult to imagine how new working and living environments could be created on the basis of new inventions in building. For example, Paolo Solery designed enormous habitats (hyperstructures) of extremely high human population density, called 'arcologies' (Figure 1)[85]. A combination of architecture and ecology, Arcologies contain a variety of residential and commercial facilities with the purpose of minimizing individual human environmental impact. The concept probably approaches most closely what urban planners for the last decades had in mind for our cities. Such integrated megacities would greatly increase the efficiency of

85 http://www.arcosanti.org/

land use, transportation and energy. They would truly transform society and even create opportunities building settlements on the moon or Mars.

More recently, the concept of the 'vertical farm' was introduced. In vertical farming, agriculture - always a typical countryside phenomenon - moves into cities, growing crops in tall, specially constructed buildings. Growing crops in this way would take advantage of hydroponic and aeroponic technologies, which are soil-free and would allow growing plants in a water-and-nutrient solution or in a nutrient-laden mist, respectively[86].

Apart from some localized experiments, such as Arcosanti, an experimental town that began construction in 1970 in central Arizona based on the aforementioned Paolo Solery's work[87] and some planned city projects, such as Masdar in Abu Dhabi[88], nothing has changed since the revolutionary suburbs of the first half of the 20th century. Initially, suburbs were comfortably connected to the cities by train and somewhat later by that quintessential American invention of the parkway. But with the suburbs morphing into the exurbs, transport has never been able to keep up. For public transport that is not surprising since by definition it serves the needs of the collectivity, not the need of the individual. Unfortunately for the individual, however, private transport has also been suffering with the once beautiful parkways often resembling large parking lots during rush hour. But of course, lots of people prefer to live in the countryside rather than dense cities, at least something that in their imagination looks like the countryside and is not too uncomfortable. In fact, they like classical houses and accept the inconvenience of traffic jams to get to work.

Meanwhile, significant improvements of our inner cities and suburbs alike have led to a situation where most of the inconvenience is tolerable and life pleasant and safe. My wife and I live in an apartment in New York's Harlem neighborhood, not long ago considered as one of the most dangerous places in the US. We sometimes go for dinner to a restaurant on Broadway, which is on walking distance. From our apartment on Frederick Douglas Boulevard (8th Avenue), the shortest way is to cross Morningside Park, which I thought was a real gem of a park. Crossing it and climbing up the stairs to Columbia University gives you a beautiful view and quick access to Broadway and some of the nice restaurants there. Quite often we

86 http://www.verticalfarm.com/
87 http://www.arcosanti.org/
88 http://www.masdarcity.ae/en/

came back late and went back through the park when it was already dark. Sometime last year I had dinner with Charlie Cantor, a famous molecular geneticist and one of the fathers of the human genome project. He used to teach at Columbia and is a real native son of New York but has lived in San Diego for a long time. I still remember how amazed he was when he heard that I lived in Harlem and dared to cross Morningside Park, in his days a place of drug addicts and violence. Now, people are playing ball and having barbecue. Designed by the famous Frederick Olmsted, who also designed Central Park, the city accorded Morningside Park landmark status in 2008, so late because of its bad reputation.

Improvements of cities and not only cities but also suburbs and exurbs is called 'gentrification' and it is a very positive development. However, it is inherently conservative and hostile to the kind of transformative changes so common in the past. For example, it does not seem to make sense that in the 21st century cities are paralyzed when the heavens decide there will be snow. Couldn't we use electric heaters under the street and pavements to avoid this endless hassle of snow plowing, digging your car out or even being blocked from getting to work? In a sense, the similarity of current city life to that of a generation ago (even the streetcar has returned, although it is now called 'light rail') may be the clearest example of how disappointingly slow technology development has become. But it did not need to be this way.

A major technological development that would really change city life would be the introduction of automated tunneling systems. More than a century ago it was obvious that the best place to put transportation in cities is underground. The tunnels for the original New York subway in the early 1900s were created manually by blasting. For the Second Avenue line currently underway, a tunnel-boring machine is used, but there is a lot more to be done to make it ready for the first train rides. Therefore, the cost of criss-crossing cities with tunnels is high. But designs have been made for boring machines that not only do the drilling, but also evacuation of the cuttings, shoring walls and ceiling, pouring of concrete, putting the ground support installation in and constructing the electrical and ventilation systems. Widespread application of such integrated systems would lead to further innovation and enormous cost reductions in tunnel construction. It would undoubtedly change the face of our cities dramatically.

During my occasional visits to my hometown of Rotterdam, The Netherlands, I am always reminded of a not so remote past when our priorities in big city design were not limited to pedestrian plazas, bike paths and waterfront promenades. Its 1931 Van Nelle factory remains a futuristic model for how current cities could and perhaps should have been[89]. Rather than moving the factory out of the city, a process which by now has been completed for virtually all big cities, the idea here was clearly to create a light, airy and spacious building in an attempt to make industry an integral, high-quality component of daily city life. Examples for such urban factories were not at all unusual in those days and we find them in many other places.

But it was not to be! Instead, we decided that the dazzling spectacle of the city of tomorrow was certainly not for today and we are stuck with all these boring, but oh so comfortable and safe neighborhoods. The point here is not to say that all these ideas are practical or cost-effective. I just want to point out that our big cities changed so little over time not because of a lack of ideas or opportunities. As with all inventions, most undoubtedly would never make it to the real world, but the real issue here is that none of them even got a chance. Our stalling city life is a consequence of a conservative streak that began to dominate our world half a century ago. Before we analyze this in some more detail, we will first discuss some seemingly simple alternative explanations for a technology slowdown.

Is there a natural end to progress?

The first explanation that is usually proposed for a possible technology slowdown is that technological progress has a natural end, which we may have reached. Jonathan Huebner, whom we met in Chapter 1, considered this the most likely explanation for the declining rate of innovation detected by him on the basis of the decrease in major inventions and patents, corrected for population growth.

In discussing a possible natural end to progress, we should distinguish physical from economical constraints. To begin with the first possibility, it goes without saying that any kind of progress that violates the known laws of physics is impossible. And some may argue that we now begin to hit some iron walls constraining further progress in certain areas. However, according to Michio Kaku in his book *Physics of the Impossible* there are surpris-

89 http://www.mimoa.eu/projects/Netherlands/Rotterdam/Van%20Nelle%20 factory.

ingly few such physical limitations[90]. Many of the limitations that have been brought up appeared later to be imaginary. Lord Kelvin's famous statement that 'heavier-than-air flying machines are impossible' comes to mind, and I remember a Dutch chess grandmaster, Jan Hein Donner, who maintained that a computer could never beat him (he was also convinced that women could not play chess!). He died a few years before Deep Blue in 1997 beat Garry Kasparov, the world's best chess player.

Some isolated areas of technology development may have dead ends, such as, for example, the horse collar, invented by the Chinese in the 5th century (Chapter 8) or the paper clip, invented in 1867 in Europe. In most if not all major areas of technology development we certainly have not reached such a stage. As we have seen in the previous chapters, there really is no physical reason why smart highways could not by now have transformed the Eisenhower Expressway, just like the motorcar took over from coach travel early in the 20th century. Or why supersonic and hypersonic passenger planes could not have taken over from the jetliners of the 1970s, just like these jetliners took over from the propeller planes in the 1950s. Or why atomic energy could not have replaced fossil fuels in the 20th century. If technological progress really goes on unabated there should by now be plenty of concrete signs to make our society visually different from the 1970s in the same way as the 1970s were different from the 1930s and the 1930s were different from the 1890s. This is certainly physically possible. Yet, it didn't happen!

So, if it is not the law of physics per se that constrains us, could it then be that inherent limitations of the human brain keep us from expanding our technological legacy? Is it possible that technologically we have picked all the low-hanging fruit and are now mired in generating progress in areas that are utterly beyond our capability? One could easily draw that conclusion from the stalemates in, for example, drug development, finding new energy sources and rocket technology, all areas using up enormous amounts of resources in terms of cash and human energy, with disappointing results. Is the idea wrong that human inventiveness, stimulated by modern markets, will always trump problems, such as scarcity of natural resources? Are there technical problems that even the best scientific minds cannot crack? Will we never have batteries that last forever, and are there no ways to tap everlasting energy sources? Is it impossible to build rockets that can escape the

90 Kaku M. *Physics of the Impossible*. Anchor, New York, 2008.

earth's gravity for less than $1,000 a pound? In other words, is the capacity of our species to innovate when and where this is needed, not infinite?

Of course it is by definition unknowable if the human brain is indeed limited and simply incapable of tackling certain challenges. Yet, there is no scientific evidence for a physical limit to human brain power. The brain may contain as many as 100 billion neurons exchanging signals among each other through as many as a quadrillion little gaps called synapses. The human brain has not changed since our days as hunter-gatherers, but now needs to cope with an increasingly complex world and a greatly extended range of tasks. One could, therefore, argue that because our brains were never designed for such gigantic loads of information as they are currently experiencing human brain power begins to limit technological progress. It is mainly for this reason that Ray Kurzweil, the main proponent of the singularity theory as discussed earlier in this book, assumed the need for machine intelligence to eventually surpass human intelligence to reach that Nirvana of ultra-high levels of intelligence.

While it can never be ruled out, limitations of the human brain are highly unlikely to constrain further technological progress, certainly not on the short term. While it is true that the brain has not changed since the origin of the human species, it is not a computer that needs programming and frequent upgrades with faster chips. Learning of complex tasks such as speaking, writing, math, music, etc., stimulate the brain cells to form more and more branches (dendrites) for signaling among them. In this sense, the brain is not immutable but highly flexible and subject to modification when it receives input. While we do not know how this connection between body and mind exactly works, we can at least say that it is far from sure that our own intelligence is now failing us and we are in desperate need of machine intelligence, just to keep technological progress going. And again, many of the solutions to ongoing technological progress are already there, as we have seen in the previous chapters.

It should also be noted that while technological progress may be on the decline, this is certainly not true for progress in science. At the same time innovation stagnated, ample progress has been made in science. While science is not the same as technology development, it does show that we are still very good in using our brains in generating important new findings, dramatically increasing our understanding of the world around us. Science now offers considerably more leads for technological innovation than it has ever done and there is no reason, at least not a scientific reason that these leads could not be followed up and result in practically applicable tools.

If there are few physical constraints to human creativity, could technological progress suffer from the law of diminishing returns? That is, could there be economic limits to technology development? This could explain why so many products of our creativity no longer reach the marketplace. It could also explain why so many of us believe that technological progress is increasing ever faster, which as we have seen is almost entirely based on progress (real or imagined) in information technology. Such people may argue that technology did progress, but not in all areas equally dramatically and also not as visibly. While in the past progress was haphazard, so the argument goes, nowadays we know exactly what we are doing and we choose the areas that need progress very carefully. We simply innovate selectively only in those areas where progress matters. It may be true that the last 3–4 decades witnessed little change in highly visible areas, such as aeronautics and space, energy and medicine, but why should we worry about the lack of a better rocket when we can call anyone on the planet for almost nothing while walking down the street? Or watch a movie on our iPad whenever we feel like it? As the argument goes, in select areas, i.e., information technology, there has been progress and it is these areas that matter because all others were already doing fine. In other words, we have become smart enough to prioritize the areas of technology that need improvements and it is this prioritized progress that has globalized and digitized our world.

Whether or not we agree that in information technology progress still is exponential, it is simply impossible to argue for a saturation of demand in all the areas where progress is lacking. One may disagree on the need to go to the moon. But there cannot be a difference of opinion about ending the traffic jams or avoiding 14-hour flights in cramped seats. And can there be any doubt that new, clean and powerful sources of energy, such as nuclear fusion, would benefit humanity enormously? The current snail's pace at which carbon as by-product of energy is reduced has become a very serious environmental threat in the eyes of many of us. Would a rational planner really give priority to yet another electronic gadget or even a valuable but not urgent improvement in computer capacity? And then there are the irritating flight delays as soon as there are a few snowflakes on the horizon. Wouldn't it be desirable to resolve this problem, for example, by using infrared heating systems for de-icing positioned over the aircraft while this is parked at the (covered) gate? And is it really necessary to still assemble and stitch garments by hand, i.e., using the sewing machine of 1830? Are we doing that because sweatshops are good for the economy of developing nations? Clearly, there is opportunity for innovation in many areas.

It is also not true that the technological advances that would resolve these problems are always economically unfeasible. In fact, British Airways operated the Concorde at a profit and given the time further innovation would have turned supersonic or hypersonic passenger transport into a routine product, similar to the current jetliners.

It seems highly unlikely that humanity has reached a stage where we have become smart enough to prioritize the areas of technology that need improvements. It is also highly doubtful that such a prioritization is possible, because inventions cannot be planned. Humans invent by default. In the right environment and given the opportunity, they invent and pursue potential applications, often even in the absence of rational arguments. Eventually, many of them fail but inevitably many others turn out to be useful in a way that could not always be foreseen. No, instead of physical or economic limits to technology development, the answer to the question of where the current technology slowdown comes from is our very success in organizing our society. Let us now take a closer look at that seeming paradox.

Vested interests

The possibility that changes in society have made it more difficult to implement new technology now than in the past seems counterintuitive. Indeed, the need for innovation is now generally recognized, which was not always the case. There is also much more money available for research and development. However, it has been recognized for quite a while that there are major factors in societies that constrain inventions from penetrating society because they pose a threat to established interests. Established producers have incentives to block innovations because they often destroy their market position. When such old industrial leaders are able to lobby governments into raising barriers to change, technological progress could easily be thwarted. For example, in the US, efforts to take drastic measures to improve energy efficiency and further develop renewable sources of energy have been seriously constrained by vested interests, e.g., fossil fuel producers, electricity providers, car manufacturers. Representing the American automobile industry, John Dingell, congressman for Michigan and member of the Democratic Party, who once headed the powerful Energy and Commerce Committee, fought tooth and claw to block nearly every regulation you can imagine, be it on air bags, tailpipe emissions or gas mileage.

The threat that innovation can stall because of vested interests is far from new. Indeed, the rise of interest groups blocking further progress is the main theme of Mancur Olson's *The Rise and Decline of Nations*, published in 1982[91]. We always had that kind of conflicts between pioneers and vested interests and it may have delayed but did not block widespread applications of the steam engine, the train and motorcar or the steamship, electricity or computers. It often took a long time to get technologies accepted that are now taken for granted. Major inventions, such as the telegraph, telephone and computer were not accepted on day 1 and took intensive lobbying to secure their implementation. But are there no persistent, eloquent inventors anymore to make grand claims about how their brainchild will change the world? Are there no politicians or business people left who believe these claims and spend their money where their mouth is? There is no doubt that such people are still around. Examples are the aforementioned Elon Musk, who spent more than 100 million of his own money (he made his fortune through the PayPal online payment service) to build the Falcon 1 launch vehicle. Why then are we so much less successful as in the days of Stephenson, Morse, Pasteur and Fleming? To understand this, we need to realize that inventors and their political and financial support groups do not work in a vacuum. Success or failure of an invention is not solely determined by these players, but by the leverage that society allows them to have. Therefore, to understand our current technology slowdown we need to know more about how human society can impact on innovation and if something changed over the last decades.

A victim of our own success

Ironically, the current decline in innovativeness plays itself out against the backdrop of an increasingly successful society. And this is not limited to the US or Europe, but a worldwide phenomenon[92]. While great human tragedies like Darfur, Bosnia and Rwanda will always dominate the news,

91 Olsen M. *The Rise and Decline of Nations*. Yale University Press, 1982.

92 Goklany I. "The improving state of the world: why we're living longer, healthier, more comfortable lives on a cleaner planet". Cato Institute, Washington DC, 2007.

most of the world is living in peace, with the number of conflicts steadily decreasing. This is undoubtedly due to a growing tendency to solve differing worldviews and conflicts by negotiation and sometimes international intervention. As a consequence, the number of refugees is falling and the human condition improving. Life expectancy at birth of the world's population has risen continuously from 48 years in 1955 to an anticipated 66-97 years (depending on the country) in 2050, according to the United Nations[93]. Mortality is declining and fertility rates are decreasing. While still increasing, world population is now expected to peak at about 9 billion in 2075 and then remain stable at replacement level. According to UNESCO, illiteracy rates have fallen from 37% of all people over 15 in 1970 to less than 18% today[94]. Also the number of children without primary education is rapidly falling. While income disparities are still enormous, world poverty is on the decline[95]. And it goes without saying that the number of internet connections is increasing dramatically, offering previously unheard of opportunities for people in all parts of the world.

The cause of all this good news is technological progress, which drives increased productivity and economic growth. In turn, this is a product of the free flow of ideas and information and an attitude of healthy skepticism that invites logical thinking and experimentation, undisturbed by hostile governments or an intolerant public. While technological progress still betters society, for example, by providing the rural poor with cell phones and off-grid solar systems (Chapter 2), the great improvements in developing nations are now mainly a result of increasingly good governance and improved ethical standards. Indeed, as we will see it is this positive development in the human condition that invites the technology slowdown that is the topic of this book.

Recently, there has been concern that science and technology are weakening, eroding the world's potential for further economic growth. In a much publicized report from 2005, *Rising Above the Gathering Storm*, prepared for the US National Academy of Sciences, a panel of research leaders expressed such concern specifically for the US[96]. Reconvening five years later, the

93 http://www.un.org/esa/population/publications/longrange2/WorldPop2300final.pdf

94 http://www.uis.unesco.org/en/stats/statistics/literacy2000.htm

95 http://web.worldbank.org/WBSITE/EXTERNAL/TOPICS/EXTPOVERTY/0,,contentMDK:20195240~pagePK:148956~piPK:216618~theSitePK:336992,00.html

96 Rising Above the Gathering Storm. www.nationalacademies.org.

panel concluded that the situation had not improved but had worsened[97]. Just to mention some of the conclusions: the federal government funding of R&D as a fraction of GDP had declined by 60% in 40 years, US firms spend much more money on litigation than research, no new nuclear plants have been built in the US over the last decades, more arts majors graduate nowadays than engineers, after the retirement of the Space Shuttle in 2011 the US relies on Russia to send astronauts to the international space station, and the number of new drugs approved by the FDA has dropped to about half since the 1990s. Clearly we have an innovation problem!

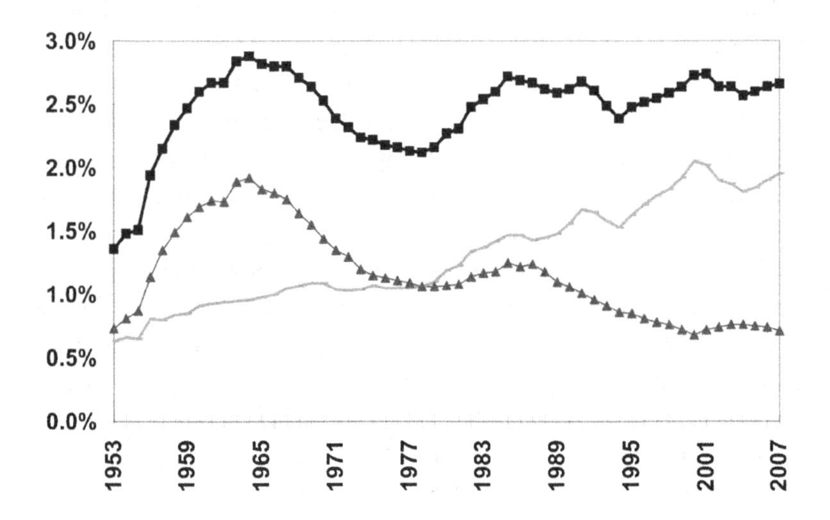

Fig. 6.2. US spending on research and development since 1953 as a percentage of GDP (gross domestic product). Source: National Science Foundation. Black square: total US R&D, grey triangle: total federal R&D, grey line: non-federal (industry) R&D.

The nature of the problem we are facing is two-fold. First, there is a lack of federal support for basic research (Figure 2). In the past, generous support in the US from such federal organizations as the National Science Foundation, the National Institutes of Health and the Defense Advanced Research Projects made it possible for talented individuals to create the foundations for many breakthrough inventions. Nowadays there is not only less money but there are also new strings attached to the grants that are given out. Over the last couple of decades basic research, especially in the natural sciences,

97 Rising Above the Gathering Storm, Revisited: Rapidly Approaching Category 5. www.nationalacademies.org

ceased to be valued for its own sake. There are increasing demands for applicability and a new focus on intellectual property protection. Somewhat surprisingly, this is not entirely dictated by the funding agencies but seems to be something the universities have done to themselves, possibly with an eye on raising revenues. Every University now has its office of technology transfer, office of biotechnology or office of business development. While in the past, as I can testify, every investigator simply sent his/her reagents to every colleague who asked for it, nowadays you need a so-called material transfer agreement. This takes time and is a hindrance to progress.

This tendency to focus on practical application is likely also the reason that industrial support for university research is on the decline[98]. Clearly, rather than risk the dollars on basic research, which may be for naught anyway, they are better spent on a more clearly defined, more likely to reach goal. While the reader, as taxpayer and possible company shareholder may generally appreciate this focus on applicability and feels satisfied that his/her tax/investment dollars are no longer left to the imagination of the average investigator, we need to realize that almost no breakthrough inventions can be planned ahead of time. Nobody could know that the first Wright flyer would give us the airline industry and we could not know that playing with bacteria would be the foundation of the biotechnology industry. Rather than living in a dream world, believing that revolutionary new technology can be planned and managed, we need to realize that our entire world as we know it, with its energy, transportation and information/communication infrastructure is a product of unfettered, curiosity-driven research[99].

The second cause of our innovation problem involves industry. In contrast to government-sponsored, basic research, corporate research has not declined and has remained rather stable or even increased (Figure 2). However, in contrast to the situation in the past when industry often conducted daring research and came up - apart from a great many failures - with some fantastic inventions, such as the transistor, the mobile phone, graphical user interface and the laser, most current research is focused on short term applied work. The great corporate labs of yore, such as AT&T's Bell Labs, where the transistor, mobile phone and UNIX computer operating system were invented; Xerox's Palo Alto Research Center (PARC), whose

98 Rapoport AI. "InfoBrief", National Science Foundation, September, 2006.
99 Buxton B. "The price of forgoing basic research". *Bloomberg Businessweek*, December 17, 2008.

scientists invented the graphical user interface and ethernet; and RCA Labs, with such macro-inventions as color television and liquid crystal displays, are no more or at least no longer the same as they used to be. Nowadays our idea of an innovative company is Apple, which, as we discussed earlier, relies much on previous inventions made by others. This is not to downplay Apple's contributions to the innovation process. To bridge the gap between an invention and the store shelf is critically important. Unfortunately, without breakthrough inventions or the courage to bring them to market, innovation itself is no longer as exciting or transforming. After all, the technology behind the iPhone is really nothing new, nor is the instrument transformative. Current corporate research is very good in what they are doing in innovation, but they no longer make breakthrough inventions. For that, we need to go back to earlier, wilder times.

Innovation problems are absolutely not restricted to the US, albeit it is now the height of fashion to compare ourselves (unfavorably) with China, which has become the new Japan. However, neither China nor Japan is doing any better than the US (or Europe for that matter). At least in the US there was and still is real innovation. China's leaders are now working hard on a top-down approach to make their state more innovative and less imitative. Their strategy is to greatly increase the number of patents that its residents and companies file by an array of incentives, such as cash bonuses and tax breaks. Unfortunately, invention and innovation are more suited to an unruly, imaginative and somewhat chaotic society like the US. Japan never gained stature as a particularly innovative country because its society abhors unruliness even more than China's communist leaders. Naturally, the point I want to make here is that instead of Europe and Asia becoming more like the US, it is the latter that has mellowed down quite a lot since its heydays of innovation.

If our technology slowdown is ultimately caused by a certain complacency that has begun to infiltrate even previously highly dynamic nations, such as the US, what are its proximate causes? That is, what spurs the current deceleration of breakthrough inventions after approximately 1970? After all, it has nothing to do with insufficient brain power, a decline in talented individuals or diminishing needs.

One direct cause is undoubtedly the declining interest in science and technology and a tendency to focus on careers in more financially rewarding areas. A shift of students away from science and engineering towards areas, such as finance and business, art and entertainment, is understandable. Indeed, pursuing a career in the natural sciences requires many years of hard work and intense study, including mathematics (calculus, probability and statistics), modern physics, chemistry and biology. Society tends no longer to reward these efforts as it did in the past. Indeed, many engineering graduates in the US are taking jobs on Wall Street to work in the finance sector. They use their math and computer modeling skills to create elaborate financial models and earn many times the pay of an engineer who merely builds useful products for society.

Even in those cases where science and engineering have been able to capture interest, we see an increasing tendency to seek a career in making gadgets rather than pursuing, for example, the development of new, alternative sources of energy. Electronic gadgets have become a major focus of our information age. All over the world we see people, both adults and children, with earphones on and little devices in their hand. There is no doubt that all these IT gadgets are based on extensive innovation, but none represents a real macro-invention. Nevertheless, it is hugely profitable and not subject to major regulatory constraints. Hence, it attracts talented scientists and engineers away from other areas of technology development.

I was reminded of the power of the gadget when reading the Korean Times during one of my visits to the 'land of the morning calm'. South Korea is a somewhat exceptional Asian country in the sense that its people are far from 'calm' but instead rather passionate and the country is more chaotic than its two neighboring industrial giants: China and Japan. In fact, Korea reminds me most of Italy. The country is making some serious attempts to be innovative and tries to come up with what it believes could be the technology of tomorrow.

The story I was reading in the Korean Times was about Yu Yeonsik, the CEO of Digital Cube Co., a Korean company that developed the i-Station V43, a portable music player that also allows viewing slide shows and we are talking a number of years ago, well before the iPhone. Yeonsik received a doctorate from the prestigious Seoul National University and subsequently worked for 4 years as a researcher at the US Fermi National Accelerator Laboratory where talented physicists study quantum mechanics. With this background and an interest in tech gadgets, he founded Digital Cube with two others in 1999. The question, of course is why a talented individual like Yeonsik did not go on developing the next miniaturized nuclear fusion reactor, but instead got himself involved in making gadgets. While there can be no doubt about the importance of the advances in information technology that are now giving us all those great tools to further our desires to be entertained non-stop at every conceivable place, one could wonder about the breakthroughs in significant areas that now escape us because too many talented individuals decide to pursue their career in areas less subject to controversy and regulation, and more profitable.

Alas, society is no longer paying high rewards for developing new, breakthrough technology, but is instead focused on what is already there. It seeks new and improved applications of existing technology in areas where they can expect to profit optimally. Such new applications require new business models for marketing and financing, with some tweaking of existing technology. This development logically follows from an increasingly sophisticated society where it simply makes more sense to invest in existing technology than in something completely novel. In the 19th and early 20th century this was not an issue because the fruits of the industrial revolution had not yet matured and the world was still a rather chaotic place with large areas of the globe closed to the world economy. Industry is simply doing what it does best and that is selecting what is most likely profitable the quickest. This goes at the cost of new, breakthrough inventions.

In the previous chapters we have already seen that there are plenty of areas where breakthrough inventions remain to be made. Here are some others. First, let's take a look at the plastic car. Modern plastics are sturdy enough to replace molten metal in airplanes or cars. Yet, car makers have always refused to consider its benefits — major weight and cost savings — to be worth the risk of doing something new. Most people would argue that this is because the technology is just not there yet. However, Henry Ford himself already experimented with a plastic car and the first plastic engine appeared in the 1970s. As always, our sophisticated business models see disadvantages that in the old days would surely have been overlooked. For example, high fiber costs and slow manufacturing output are real problems that would indeed make it difficult for plastic engines to immediately compete with metal ones. But wasn't it always like that? Didn't it take a long time before the steam turbine had out-competed the water turbine? If 21st century business models could have been applied in the 19th century we would surely still get our energy from water.

Let's look at a second example. Garments are still assembled and stitched by hand, i.e., using the sewing machine. Are the problems in automating the stitching of clothes and shoes really insurmountable? One could imagine, for example, an entirely new type of textile industry in which lightweight clothes are made from new materials using nanotechnology at a fraction of their current costs. Rather than wearing them for longer periods of time, apparel could become disposable and ordered every morning through the internet. A 3-D body scanner would create custom-fit clothing patterns, obviating the need for actual fitting. You select style, fabric and design from a clothing manufacturer on the Internet and e-mail your body scan. Your custom-fitted garment would be transported through pneumatic tubes. This would also obviate the need to carry luggage, resolving a major headache of the airline industry.

Intriguingly, such a scenario of revolutionizing the textile industry yet again, as in the 19th century, is not hindered in any way by either a lack of imagination or the reality. In his futuristic science fiction comedy film, *Sleeper* (1973) Woody Allen is getting fitted for a suit by two robotic tailors more or less in the way described above. In a 2004 paper Susan Ashdown described the potential of 3-D scans in fitting apparel[100]. Hence, the question is not why we are unable to further technological progress in the textile industry, but why new ideas and tools, such as 3-D scanning, are not already widely used for custom-fitting clothes and shoes. When I raised this possibility during discussions with some female friends they vehemently opposed the idea, mainly because they felt that it would take away their freedom of selecting their own clothes. I was simply unable to convince them that the internet would expand rather than narrow their choices, making dressing well as personal as can be. Nevertheless, I do agree with them that such a change in our society is not going to happen. So, we will be stuck with sweatshops and lost luggage for a long time to come.

Industry is reluctant to invest in new, breakthrough technology not only because society is conservative but also because of the severe penalties current inventors often need to pay to see their discoveries implemented. Much more than in the past, today's innovators must contend with government regulations, prohibitions on federal, state and community level and public anger, which often translates into illegal actions tolerated by the authori-

100 Ashdown SP, Loker S, Schoenfelder K, Lyman-Clarke L. "Using 3D scans for fit analysis". *J. of Textile and Apparel Technology and Management* 4, 2004.

ties. This can best be illustrated by an example from an area of technology development not yet discussed: agriculture.

In farming, the best results were usually obtained in the absence of meddling by governments or administrators. We see this time and again with the best examples the failed collectivization of agriculture in the former Soviet Union. Indeed, agricultural miracles occurred when farmers were free to do as they liked with the opportunity to optimally profit from technological progress. In this way, we have seen the rise of the iron plow, fertilizer, seed drill, tractor and pesticides, to mention some major developments. As we have seen, it was great progress in agricultural productivity that prevented the predicted Malthusian crisis in the 19th century. Dramatic improvements did not stop in the 19th century. In the 1940s and 50s world food production more than doubled because of the development of high-yielding varieties of disease-resistant wheat. The name of Norman Borlaug (1914-2009) will always be associated with this agricultural miracle.

Of course, as part of its success society has come to regulate most aspects of our lives including agriculture. And this is a good thing in view of the demonstrated risks of pesticides and other harmful agricultural practices. However, with the introduction of the latest feather in the cap of the productive agriculturist, i.e., genetically manipulated food, it seems that the end of further progress has come. As with other areas of technology development, large sections of the public no longer conform to the advice of experts, but decide for themselves what is right and what is wrong. This results in exceedingly lengthy time periods of regulatory decision-making, an important constraint on the ability of industry to innovate and generate new products. For example, the time it took regulatory agencies to reach decisions on genetically modified crops more than doubled between 1994 and 2005, with no explanation given[101]. These delays are more likely caused by pressure groups than driven by scientifically rational arguments. Indeed, while genetic engineering still instills angst in the lay public, strongly fueled by some fanatic environmentalists[102], there is no scientific reason to worry about safety of GM foods, which can greatly benefit farmers, consumers and

101 Jaffe G. "Regulatory slowdown on GM crop decisions". *Nat Biotechnol* 24,748–749, 2006.

102 Schiermeier Q." German universities bow to public pressure over GM crops". *Nature* 453, 263, 2008.

the environment[103]. Nevertheless, logic rarely wins in debates on safety or desirability of implementing new technology.

Regulatory problems even occur after a GM crop has been widely adopted by farmers. For example, in September of 2009, a federal judge in San Francisco ruled that the government failed to adequately assess the environmental impacts of genetically engineered sugar beets before approving the crop for cultivation in the US (where the situation is generally a lot better than elsewhere)[104]. This decision, which could lead to a ban on the planting of the beets, is very similar to a ruling two years earlier that farmers could no longer plant genetically modified alfalfa. The agriculture secretary has now pulled back that decision and the alfalfa can be grown[105]. Both the sugar beets and alfalfa GM variants contain genes that render them resistant to pesticides, which will then allow farmers to kill weeds or insects without harming the crops.

This problem is not limited to the spoiled palates of Europe or the US. Opposition from activists has led the Indian government to block commercial release of that country's first GM food crop - a genetically manipulated eggplant resistant to insects - despite clearance by its biotechnology regulators[106]. Genetically modified crops could also solve very serious problems in as yet underdeveloped countries. For example, genetic modification can be used to biofortify food crops to make them rich in micronutrients. A lack of micronutrients in, for example, cassava, the staple of many an African country, causes malnutrition[107]. Nevertheless, African governments reject GM foods, undoubtedly because of regulatory niceties in developed countries where the average citizen can afford to be anti-GM even if there is no rational argument for such a stance. It is quite obvious that in this case increased regulatory scrutiny has serious consequences. Not only does it make it much harder for developing nations to reach the stage of food security that Western countries reached more than a century ago, it will also prevent further price declines of food. Indeed, the relative cost of basic food-

103 Arntzen CJ et al. "GM crops: science, politics and communication". *Nature Reviews Genetics* 4, 839-843, 2003.

104 Pollack A. "Judge Revokes Approval of Modified Sugar Beets". *The New York Times*, August 13, 2010.

105 Pollack A. "U.S. Approves Genetically Modified Alfalfa". *The New York Times*, January 27, 2011.

106 "India halts release of GM aubergine". *The Guardian*, February 9, 2010.

107 Kristof ND. "Bless the Orange Sweet Potato". *The New York Times*, November 24, 2010.

stuff in the developed world has dramatically decreased as a consequence of new technology, but is now gradually increasing everywhere.

Irrational arguments, easily refuted by some experts, are now suppressing technological progress everywhere. Yet another example is the debate about human embryonic stem cell research. As described in Chapter 4, such cells, which can differentiate in essentially all cell types in the human body, have great potential in medicine[108]. However, ethical objections to the use of human embryos for obtaining embryonic stem cells greatly hamper progress in this area.

Examples of this retrenchment abound and even extend to technological developments that are hardly dramatic and merely involve rather benign shifts. In principle, as we have seen in Chapter 2, wind- and solar-based generators can provide for all our electricity, thereby obviating the need for greenhouse gas-generating fossil fuels. However, to efficiently mobilize those sources we need a bigger and smarter grid. Indeed, our current electric grid is very similar if not essentially identical to the one first switched on by Edison in 1882 in New York. With the emergence of highly efficient wind turbines, 100% clean energy seems right around the corner since building lots of wind turbines and expanding the grid is pretty straightforward. Unfortunately, similar to the situation with GM crops and stem cells, there is some heavy opposition not only to the actual construction of wind turbine parks, but also to expansion of the grid. For a line that would take two months to build sometimes many years of regulatory approval go by[109]. In one of the most blatant attempts to blunt a major step forward in clean energy generation, the first US offshore windmill park was delayed by many years because of a variety of flawed arguments, varying from bird migration to shipping dangers[110].

Irrational resistance by the public, the tolerance of governments to illegal actions and personal threats by activists and a crushing load of regulations reflect a desire for stability that does not provide much incentive for daring innovations. Compare that with the early steam railroads, which advanced relentlessly in spite of extensive opposition, not all based on groundless fears. In our own times, the owners of stage coaches and canals

108 http://stemcells.nih.gov/info/health.asp

109 Wald ML. "Hurdles (Not Financial Ones) Await Electric Grid Update". *The New York Times*, February 6, 2009.

110 "Blowhards". *The Economist*, November 9, 2009; Cape Wind. Editorial. *The New York Times*, November 1, 2009.

would undoubtedly have managed to let Congress pass rail safety and passenger rights bills to strangle the railroads. The current easy bowing of the authorities to public pressure (often against all scientific and legal logic) strongly stacks the odds against technology in its attempts to find applications in society. This can have serious consequences. To stay with GM foods, the readiness of authorities and public alike to adopt perfectly safe, genetically manipulated foods essentially constrains another agricultural revolution similar in magnitude to the one Norman Borlaug pioneered.

Fear of change, real change, was absent from the minds of our forebears 100 years ago. This was again shown to me when I recently visited the exhibition *The Great Upheaval* in the Guggenheim museum. The artists of the early twentieth century positively craved change. The group of artists formed by Vasily Kandinsky and Franz Marc in 1911, which they called Der Blaue Reiter (The Blue Rider), preached a 'große Umwälzung' (great upheaval) that would radically change the art world. Filippo Tommaso Marinetti, Umberto Boccioni and others of the Futurist movement, celebrated speed and the machine. They rejected the past and embraced all things modern, an attitude that characterized the tremendously creative and innovative, yet relentlessly violent first decades of the twentieth century. In many respects our current world was created by them.

Not that there was no opposition to the implementation of novel technology in those days. There have always been vested interests threatened by something new, such as stage-coach owners whose occupation was replaced by the railroads. The difference between then and now is that virtually all activities that take place in our current society are regulated. And let there be no misunderstanding; the great increase in regulation since those violent times to control human behavior has improved society dramatically and contributed greatly to our well-being. We wouldn't want to miss it! Yet, there is a price to be paid in the form of diminished incentives to develop and implement new inventions[111]. Some of this is over the top. Do we really need to live with arguments that touch close upon the absurd, just because they are nevertheless accepted by the public and tolerated by most governments, all because of the desire for stability? Under such conditions it is far from surprising that our most talented people prefer to get rich in the safe world of entertainment or finance. In the Epilogue I will discuss our options

111 Petkantchin V. "Risks and regulatory obstacles for innovating companies in Europe". Institut économique Molinari, Bruxelles, 2008.

for mitigating some of the most adverse effects of regulation. But don't hold your breath; regulated societies are conservative by their very nature and it remains to be seen if regulation can be reconciled with innovation.

Finally, before I summarize this part of the book, we need to discuss two issues that some may consider as factors in the tendency of society to no longer quickly adopt potentially destabilizing inventions. Reduced risk taking is often associated with older people or women. Older people are generally considered wiser and less inclined to do foolish things than young men. Similarly, some would argue that women also are less inclined to take a risk than men. I have personally no real opinion on the matter and as a non-expert I am unfamiliar with the underlying literature to support these ideas. But, for the sake of argument, let us assume these are facts.

There can be no dispute about the graying of society, and in Chapter 4 I already pointed out that due to food security, increased hygiene and advances in medical care people are now living substantially longer than at any time in history. Apart from the graying of society, there is probably also a feminization of society. Feminine attributes have come to be considered more important than masculine attributes, possibly because so many families are fatherless and far more women teach grade-school than men. Hence, many children are primarily influenced by females, which may have shifted the focus from competitiveness to empathy and support. Recent statistics reveal that from the elementary years and beyond, girls get better grades than boys and generally fare better in school[112].

Somewhat ironically, these kinds of arguments have been made in the past. For example, as we shall see in Chapter 9, Edward Gibbon, the 18th century British historian, believed that the Roman Empire declined because of the loss of its former masculine spirit. He ascribed this to the Christian religion, which had become dominant in the empire during the 3rd century. Whether or not the minds of elderly and/or females are genuinely associated with reduced risk taking in contrast to young men is an issue for sociology to resolve. It is not in contrast to what I state above and may or may not contribute to the phenomenon that is the thesis of this book: a technology slowdown driven by an increasingly strong reluctance of society to adopt novel, breakthrough inventions.

To summarize this first part of the book, In spite of many small improvements the human technology portfolio is no longer rapidly expanding.

112 Hoff Sommers C. "The War Against Boys". *Atlantic Magazine*, May 2000.

Major inventions have become rare and predictions from the 1960s have not materialized. We travel in essentially the same cars, trains and planes, rely on the same energy sources, live and work in the same cities, go to the same movie theaters and are still unable to cure cancer, heart disease or diabetes. However, there is one big difference: most of us do incredibly well as compared to those days of high expectations. There are still wars, there is still hunger and poverty and there are environmental threats, but they have all become much less severe. None of our cities is even remotely similar to the compact cities predicted in the early 1960s. However, they are now much less crime-infested, less polluted by exhaust fumes and safer than in the old days. They also harbor many more trees, pretty parks and pedestrian areas.

The cold war is over and most current states profess support for democracy, the rule of law and a safety net for everyone. Almost all of us seem to have abandoned violence as a way to make progress. Representatives of the earth's states meet regularly and although we tend to think that nothing ever comes out of these meetings, the reality is that there are probably very few citizens who do not profit from the myriad of changes that slowly but surely transform us into a planet-wide commonwealth of states with almost unrealistically good credentials. Good governance, responsible industry leaders and better citizens slowly but irrepressibly take hold of society by creating a network of rules and regulations that make it all but impossible to run roughshod over others. People around the world are becoming healthier, wealthier, better educated, more peaceful, and increasingly connected, and they are living longer.

Precisely because society is so successful, with technology maturing and all structures in place for optimally using it to the benefit of humankind, it has now become very difficult for new, breakthrough inventions to thrive. Investments are more likely in areas that have stabilized and proved their mettle, and regulatory constraints make it often very difficult for new inventions to get a fair chance of being implemented. This is especially true for medicine and biotechnology, two areas where scientific input should all but guarantee rapid progress. However, it also applies to the energy and transportation sectors where we witness a stalling of progress in strategic technologies. In spite of the hype in glossy magazines and the occasional newspaper article, to maintain the current fleet of passenger planes is simply too convenient and the days of Howard Hughes are definitely over. Surprisingly enough, this even applies to information technology where progress has helped us enormously to do the things we were already doing much quicker and in a much more convenient way than in the past. However, it does no longer break new ground and does not radically alter the way we live and work.

In a sense, we have become a victim of our own success. The intricate net we have woven to protect each and every one of us from harm, to allow a maximum of input to everyone in decision making and to patiently listen to all arguments have turned us into the most successful society the world has ever seen. But paradoxically, this very success is now beginning to hold us back from making the further great strides to again transform human society, like the tremendous achievements of the 19th and 20th centuries did before. However, while ours is the first global society, it is not the first successful society to emerge during our species' wandering the earth. During the history of the human race multiple, successful societies developed in parallel, often without knowing much about each other. If success breeds technological stasis, then we may observe similar cycles of innovative booms and busts with previous successful societies. If we could identify technology slowdowns as recurring phenomena based on similar characteristics of different societies in different times, then the paradoxical hypothesis that success breeds declining progress would gain tremendous strength. It would allow us to make multiple series of observations, each of which providing us with evidence for or against the hypothesis. This is the topic of Part Two of this book, where we will trace technology development through the ages to see if there were episodes of sudden declines in technological progress similar to the one we are currently witnessing.

Part Two — Parallel Worlds

Chapter 7. Human Origins: Riding the Waves of Technology

Technology can be defined as the tools to control our natural environment to satisfy perceived human needs. To find the roots of our technology we need to go back in time to when the first human species emerged on the plains of Africa, now several million years ago. In the first episode of the movie *2001: A Space Odyssey*, some prehistoric ape-men somewhere in Africa received the gift of technology from a black, rectangular monolith belonging to aliens in space. This transforms them into a new, exclusive form of higher-order, intelligent beings: *Homo sapiens*. The reality is not as romantic and certainly less exclusive. While it is true that no other species than *Homo sapiens* has made technology so much its own, the use of simple tools is not unique to humans. This is not unexpected because of the similarities between humans and other animals in terms of problem solving skills and intelligence. For example, Egyptian vultures break open ostrich eggs using stones, held in their beak, New Caledonian crows shape tools out of the barbed edges of Pandanus leaves and use them to extract food, and the woodpecker finch compensates for its short tongue in prying grubs out of a tree branch by using a cactus spine[113].

113 van Lawick Goodall J, van Lawick Goodall H. "Use of Tools by the Egyptian Vultures, Neophron percnopterus". *Nature* 212, 1468 – 1469, 1966; Hunt GR, et al. "Animal behaviour: Laterality in tool manufacture by crows". *Nature* 414, 707, 2001; Tebbich, S et al. "Do woodpecker finches acquire tool-use by social learning?" *Proc. R. Soc. Lond.* B. 268, 2189-2193, 2001.

Tool use is especially interesting in non-human primates because it provides important information on how early humans began to develop technology. All the great apes are able to use tools although they are not always observed to do so in the wild. As first recorded by the famous primatologist Jane Goodall in the early 1960s[114], chimpanzees both use and make tools. For example, chimpanzees can learn to use poles to escape from an enclosure. In the wild, they have been seen to construct tools from grass and twigs to extract ants from their holes or to withdraw honey from beehives. Capuchin monkeys use stones to crack nuts in the wild. In captivity these New World primates produce flaked stone artifacts that they use as cutting tools in a right-handed manner[115]. This indicates that the cognitive and biomechanical conditions for the production of stone tools are present in non-human primates.

It is now generally accepted that early human evolution took place in Africa with all modern human populations originally spread from there. This African origin of humans was already suspected by Charles Darwin, based on the close likeness of humans with great apes (chimpanzees, gorillas and orangutans), so abundant in that continent[116]. Current fossil and genetic evidence indicates that the last common ancestor of humans and chimpanzees (our closest living relative) originated in Africa about 6 million years ago, used tools and lived in social groups. The genus *Homo*[117] probably emerged 1.9 million years ago, with remains of its earliest true representative, *Homo erectus*, found in East Africa. This large hominin, also called *Homo ergastes*, spread to southern Eurasia around 1.7 million years ago. Numerous remains of this species have been found in Indonesia on the island of Java ('Java Man') and in China ('Peking Man'). *Homo erectus* may already have controlled fire, as became evident from burnt bones and wood in Zhoukoudian, China, which was dated to between 500,000 and 200,000 years ago[118]. After *Homo erectus* there were at least two more major movements of people out of Africa. The

114 Goodall J. *The Chimpanzees of Gombe: Patterns of Behavior*. Belknap Press of the Harvard University Press, Cambridge, 1986.

115 Westergaard GC. "The Stone-Tool Technology of Capuchin Monkeys: Possible Implications for the Evolution of Symbolic Communication in Hominids". *World Archaeology* 27, 1-9, 1995.

116 Darwin C. *The Descent of Man*. John Murray, London, 1871.

117 Homo is the group (genus) to which humans and several other hominins, such as Homo erectus and Homo neanderthalensis, belong. The genus is part of the family Hominidae, which in turn belongs to the order Primates.

118 Weiner S, et al. "Evidence for the use of fire at Zhoukoudian, China". *Science* 281, 251-253, 1998.

most recent one involved anatomically modern humans, i.e., *Homo sapiens*, a species with a large (1400 cc) brain that first appeared in Africa probably as early as 195,000 years ago. Brain size, which roughly corresponds with intelligence, increased dramatically as humans evolved. For example, Homo erectus had a brain size of only 1100 cc. Ultimately *Homo sapiens* would become the only representative of the genus *Homo* on our planet.

The expansion of *Homo sapiens* into Eurasia may have begun around 90,000 years ago and probably occurred in waves. Meanwhile, *Homo neanderthalensis* evolved in Europe and western Asia. It is generally believed that this was a species separate from *Homo sapiens* with an even slightly larger brain size. The early dispersal of *Homo sapiens* out of Africa may have followed a coastal route via the Horn of Africa, reaching Australia around 50,000 years ago, with subsequent expansions north and into North America. There were probably subsequent expansions into East and Central Asia and Europe (Figure 1)[119].

Hence, modern humans were the most recent in a series of geographically expanding migrations of hominins from Africa. Although eventually *Homo sapiens* became the only surviving hominin, they lived for a long time side by side with *Homo neanderthalensis*. *Homo erectus* persisted only until about 300,000 years ago, but the Neanderthals until 35,000 years ago and perhaps even a little longer.

Fig. 7.1. Human expansion from Africa.

119 Templeton AR. "Out of Africa again and again". *Nature* 416, 45-51, 2002; Finlayson C, "Biogeography and evolution of the genus Homo". *TRENDS in Ecology and Evolution* 20, 457-463, 2005.

In the movie *Cast Away*, we see Tom Hanks as a FedEx executive surviving a crash landing on a deserted island. To survive on his own he basically has to learn in a very short time what our species taught itself over thousands of years. Trying to open a coconut by beating it with a piece of rock, the makeshift hammer splits itself into several conveniently sharp fragments. His subsequent attempts to use one of these as a primitive axe to cut open the object of his desire would certainly have raised enthusiasm among our primitive ancestors of yore. Likewise, his enthusiastic shout 'I have made fire' after many failed attempts using a hand drill must have been a battle cry for those early humans who had enough of relying on hot ashes or burning wood from natural fires.

As we have already seen, technology is not even unique to primates and there is no doubt that tool-making by human species stretches back to the very early representatives of the genus *Homo*. The earliest stone tools of humans are associated with *Homo habilis*, an early ancestor of *Homo erectus*. They are termed 'Oldowan' after the Olduvai gorge in eastern Africa where they were first found. They are 2.6 million years old and very similar to the primitive stone tools made by the Capuchin monkeys described above. Such tools were mainly crude flakes hammered from siliceous rocks, such as flint. The flakes as well as the remaining 'chopper cores' were used as cutting tools and axes, very similar to what we see in the movie *Cast Away* with Tom Hanks.

It is likely that also *Homo erectus* initially used Oldowan technology. Later, they switched from Oldowan to so-called Acheulean tools, named after the place in France (St. Acheul) where a 300,000-year-old set of such stone tools was first found and associated with this species. *Homo sapiens* initially may have also produced Acheulean style tools. Acheulean tools were flaked on both sides (bifacial) and rather more sophisticated than the crudely flaked Oldowan tools. They included the first hand axe and also wooden and bone tools. The spread of such tools out of Africa is generally known as the Acheulean cultural expansion. The period is called the middle Paleolithic, prior to around 40,000 years ago[120].

The next stage in tool design is called Mousterian (after Le Moustier in France). This stone technology is generally associated with the Neander-

120 The Paleolithic is also called the Old Stone Age and corresponds with the middle Pleistocene, a geological term for an epoch that lasted from 1.8 million to 12,000 years before the present. It was followed by the Holocene, the time since the last ice age, in which we still live.

thals (120,000–35,000 years ago), a much more advanced species than *Homo erectus*, but must also have been used by *Homo sapiens*. There is important archeological evidence of technological prowess of *Homo sapiens* early on, indicating standardized tools for fishing and shell fishing as well as the first forms of art. Indeed, more than 70,000-year-old engravings, found together with other artifacts such as ornaments made out of snail shells, from South Africa's Blombos Cave indicate both anatomically and behaviorally modern humans. Even older are the advanced stone tools and personal ornaments found in Morocco and in Israel. Estimated to be over 80,000 years old, these and the ornaments from Blombos Cave indicate modern human behavior, including the capacity for abstract thinking, planning, symbolic behavior (as in painting and making ornaments) and, critically important, technological innovativeness[121].

Hence, it is highly likely that the tendency to innovate is an inherent characteristic of *Homo sapiens*. Humans would become the earth's dominant technophiles and innovative behavior is the human default response to environmental challenge. This has allowed humans to adapt to different environments, where other, related species could not do the same thing. Their love affair with technology may be the reason why humans were the only survivors of the many different hominins that once roamed the earth. We have already seen that *Homo sapiens* overlapped with its close relative *Homo neanderthalensis*, but only the latter perished during the last ice age. The simplest reason for their decline is that Neanderthals lacked the cognitive functions that would have allowed them to innovate their way out of trouble, i.e., the last ice age. From the beginning, therefore, humans were like us, with their languages and intellectual capabilities equivalent to those of anyone living today. A human baby transferred from 150,000 years ago to New York of 2011 would grow up like any other US child.

Human inventiveness proved phenomenally successful and the reason why around 12,000 years ago, near the end of the last ice age, there may have been as many as 20 million humans living in simple, hunter-gatherer

121 Henshilwood CS, Marean CW. "The origin of modern human behaviour: A review and critique of models and test implications." *Current Anthropology* 44, 627-651, 2003; Henshilwood CS et al. "Middle Stone Age shell beads from South Africa". *Science* 384, 404, 2004; Bar-Yosef Mayer DE et al. "Shells and ochre in Middle Paleolithic Qafzeh Cave, Israel: indications for modern behavior". *J Hum Evol.* 56, 307-314, 2009; Balter M. "Was North Africa the launch pad for modern human migrations?" *Science* 331, 20-23, 2011.

societies everywhere on earth. The success of their inventions was determined by environmental constraints. It paid to come up with fishing gear, hunting tools and equipment for processing food and to make clothing. By the end of the Pleistocene humans were distributed widely over the face of the earth, from Western Europe to East Asia and North America, from the arctic to South Africa, Australia and South America. From this 'level playing field' some, but by no means all, would now begin the developmental path from tribes to chiefdoms, states and empires. The key technological events, which subsequently led to writing, music, visual arts, crafts, politics, war and all the other traits that make up human culture, were plant and animal domestication.

The Neolithic revolution

The transformation in the early Holocene of human societies from foraging to agriculture constitutes the most important development for our species since its origin now almost 200,000 years ago. Wherever it occurred, this transition from the Paleolithic to the Neolithic (New Stone Age) spurred population growth and was ultimately responsible for virtually every subsequent technological milestone until deep into our own time.

Plant domestication and agriculture are likely to be outgrowths of foraging behavior. Through foraging humans contributed to plant propagation. By leaving seeds and root tips, they promoted an increasing abundance of certain plants that over time became cycles of planting (initially inadvertent but later deliberate) and harvesting[122].

In the beginning, hunting and primitive agriculture must have been practiced side by side. The great promise of the agricultural transition for the future of humankind was of course not foreseen by anybody at that time, and it is likely that Paleolithic humans were simply driven into a more heavy focus on growing crops and keeping animals by a decrease in wild animals, due to overhunting. For example, human over-hunting most likely caused the extinction of all large mammals in North America. It has also been suggested that a cool, dry interruption in the global warming trend at the end of the ice age, called Younger Dryas, led to a decline in food availability driving our ancestors to start sowing and harvesting wheat. Whatever the causes, the beginning of agriculture was not a conscious act, but simply a necessary adaptation to a changing environment. This does not

122 Pearsall DM. "From Foraging to Planting". *Science* 313, 173-174, 2006.

mean that agriculture was invented because it was needed. It had undoubtedly been invented many times before but was never adopted because there was no immediate advantage. In fact, agriculture is positively unpleasant as compared to hunting and gathering and it is highly doubtful that it offered any dietary advantages[123]. (In fact, after the adoption of agriculture, dropouts in the form of nomadic herders were frequent.)

Differences in environment, not innate differences between people (all members of *Homo sapiens* are equal), determined who would be set on the road to wealth and who would not (or far later). As pointed out by Jared Diamond in his magisterial *Guns, Germs, and Steel*, most wild animal and plant species are unsuitable for domestication and the fact that Africa, Australia and the Americas are less well endowed in this respect than Eurasia goes a long way explaining why the first civilizations arose in the latter continent[124]. Douglas Hibbs and Ola Olsson modeled biogeographical endowments (defined as the number of domesticable species of large animals and edible plants) as a function of current per capita GDP (gross domestic product divided by the number of people) in different areas of the world. The results show that in historical times biogeographical factors were the prime movers of civilization and wealth[125]. It was only at later times that the quality of political institutions (e.g., rule of law, property rights) became dominant, as they still are.

Not only were there many more valuable wild plants and animals suitable for domestication in the giant continent of Eurasia, the many East-West corridors greatly facilitated rapid dissemination of crops and domesticated animals as well as new technology, ideas and people. By contrast, the Americas and Africa depended on North–South diffusion, which is more difficult because of the strong effect of changing longitude on climate. Also the narrow corridor between North and South America and the Sahara barrier in Africa did not help the spreading of information and products.

Traditionally, the first crop cultivation is assumed to have taken place in the Middle East. Cultivation of annual cereal crops, such as wheat, barley and other grains, started about 12,000 years ago in southern Turkey and

123 Diamond, J. "The worst mistake in the history of the human race". *Discover Magazine*, May, 64-66 (1987).
124 Diamond, J. *Guns, Germs, and Steel*, Norton, NY (1997).
125 Hibbs DA Jr, Olsson O. "Geography, biogeography, and why some countries are rich and others poor". *Proc. Natl. Acad. Sci. USA* 101, 3715-3720, 2004.

northern Syria[126]. The natural occurrence of many wild cereals and an optimal climate for primitive agriculture in this area, also called the 'Fertile Crescent', must have been major factors for agriculture to arise there early. Indeed, the Fertile Crescent with its wooded, rain-catching hillsides with many wild plants and animals and soft soils was ideal territory for a beginning farmer. However, also China had its easy soils for beginning agriculturists. The loess soil region of the Yellow River basin in Northern China saw the growing of millet, a cereal crop, as early as 8,000 BC.

While it is possible that the shift from Paleolithic hunter-gatherers to Neolithic agriculturists took place earlier in Eurasia than elsewhere, it should be noted that there is still much uncertainty about early settlements and agricultural origins. As the historian Felipe Fernandez-Armesto once pointed out, many archeological finds are conveniently close to the major universities in Europe and the US east coast[127]. Because they were the first to practice archeology, a significant bias towards Europe and those places Europeans controlled and greatly favored (think about the romantic stories about Troy, the biblical heritage of Jerusalem or the birthplace of Abraham, Ur) is almost unavoidable. In fact, there is evidence that in America sweet potatoes were cultivated in what is now Peru as early as 10,000 years ago. And it is likely that in all places with population pressure and where the game had become scarce, crop culture became popular. It is not at all impossible that extensive excavations in America and Asia will change the picture of where exactly agriculture started. However, there can be little doubt that Eurasia was endowed with many more valuable plants and animals suitable for domestication than other continents.

The dog seems to have been the first domesticated animal and the only one during the Paleolithic age. During the Neolithic era this was now changing rapidly. In the Middle East, domestication of the wild sheep and goats roaming around in the Syrian hills almost certainly overlapped with the first crop cultivation in that area. Cattle were also domesticated early on, independently in the Middle East and the Indus Valley, as well as pigs (in the Middle East and China). The domestication of the horse as a means of transport was relatively late, i.e., around 2,000 BC in the southern Ural

126 Lev-Yadun S et al. "Archaeology. The cradle of agriculture". *Science* 288, 1602-1603, 2000.

127 Fernández-Armesto F. *The Americas*. Modern Library, New York, 2003.

Mountains, but it is possible that they were kept thousands of years earlier for their meat or as pack animals.

Therefore, the picture that emerges is that of a successful animal species, *Homo sapiens*, rapidly spreading out of Africa, their continent of origin, colonizing the globe and greatly increasing in numbers, running right into the Malthusian trap of overpopulation. I already mentioned the disappearance of all large land animals from the North American Great Plains due to overhunting accompanying the successful expansion of the early native Americans now less than 10,000 years ago. In those areas where agriculture was possible it began to be practiced. This does not mean that all these proud hunters were told to go out and invent how to culture crops. More likely, experiments with agriculture must have been done on a continuous basis, perhaps almost from humanity's origins. Only when they ran out of animals to hunt or easy to find food, crop culture established itself, simply because the alternative was either death or living under extreme duress.

It is difficult to overstate the importance of the transition from a migratory life in bands and tribes following the game to the sedentary lifestyle necessitated by the needs of agriculture. Indeed, it is the single, most important consequence of this new lifestyle, division of labor, which drove all subsequent developments. Although recurrent rounds of feast and famine, so characteristic for Paleolithic life, close to the margin of subsistence, would not entirely disappear until our own time, agriculture led to surpluses which could be stored and enabled people to devote time to other activities. It is division of labor that greatly expanded technology development, leading to the surge in wealth that would last to our own days. New technology, for agricultural but also for other purposes, created the great civilizations in the Middle East, India and China, and later still led to the emergence of the technological powerhouse of Western Europe.

First, agricultural lifestyle was not fully sedentary because soil exhaustion regularly required virgin land. Such new fields were created by cutting and burning forests or woodland. The wooded hill country where the tree cover prevents heavy undergrowth lent itself well for such slash-and-burn agriculture. Polished stone tools, to replace the chipped and flaked flint, were characteristic for the Neolithic period. Hard stone rather than brittle flint could be used to make the axes for cutting trees. Polished stone blades attached to a wooden handle served as simple hoes. This tool, together with plain digging sticks and the first primitive sickles to harvest seeds of cultivated grains were the sole tools of early agriculture.

The story of how the Neolithic transition led to village communities, cities, states and empires, and eventually to industrial and post-industrial society, has been told many times. I particularly enjoyed reading William McNeill's *The Rise of the West*, which in spite of its Western bias provides an almost poetic view on how civilizations arose[128]. Once agriculture was practiced routinely and had replaced most of the previous hunter-gathering activities, progress in developing new technology was rapid. This is undoubtedly related to the dramatic changes in life style associated with the Neolithic transition to agriculture. Temporary shelters such as tents (of animal skin) or caves were gradually replaced by more permanent settlements, first dugouts and sod houses, and finally the typical Middle Eastern houses built from sun-dried brick as well as wooden houses (in China and India). The earliest evidence for Neolithic villages is from the Middle East, China and India (including modern Pakistan and Bangladesh), but they appeared shortly thereafter in Africa, Europe and the Americas as well.

The agricultural tool palette quickly broadened, with the most important invention undoubtedly the plow. Perhaps invented as early as 8,000 BC it was originally a modified tree branch to scratch the surface of the soil and prepare it for planting. This first traction plow, not more than a heavy hoe or spade, was fastened behind a draft animal, to break open the ground. This led to a major improvement in agriculture, from the original slash-and-burn type to the more sophisticated practice of leaving part of the land unsown to regain fertility. Because the plow allows crop residues and weeds to be turned into the soil, plowing fallow land several times during the growing season allowed it to regain fertility. In this way, fertility could be maintained indefinitely. The first plows were made of wood and had no moldboard to turn a furrow. Later they were made with iron points, and in China an iron plow was developed in the third century BC (see Chapter 8).

The use of animal power in agriculture was a first example of how domesticated animals could interact with agriculture. In the Middle East and China (and later in Europe) oxen were the primary draft animals. In South Asia, it was the water buffalo. It was soon found out that crop yields could be greatly increased by manure applied as fertilizer. Although rice, which

128 McNeill WH. *The Rise of the West: A History of the Human Community*. University of Chicago Press, 1991 (originally published in 1963).

can be grown wet or dry, is the only subsistence crop that needs no nitrogen fertilizer, manuring and plowing of rice fields to increase yields were also done in China and India.

The first querns and mortars for grinding grain were invented at about the time agriculture evolved. Pottery everywhere replaced stone containers, probably preceded by basketry. The potter's wheel came later, i.e., around 3500 BC. In the Middle East, linen textiles were manufactured from domesticated flax and woolen textiles from domesticated sheep. In China, hemp and silk textiles were derived from the domesticated hemp plant and silk worm, respectively, and in India cotton textiles were derived from the cotton plant. Such textiles provided for the first time substitutes for the skins and furs used by Paleolithic people. The first sailing ships and wheeled carts emerged around 3200 BC. The wheel was probably invented around 3500 BC in the Middle East and first used in China and India around 2000 BC.

In the Middle East, the first metallurgy, based on copper began as early as 5500 BC, followed by bronze (an alloy of copper and tin) around 3100 BC, possibly somewhat later in China and India. Iron smelting was probably invented in the Middle East, perhaps in Anatolia (what is now Asian Turkey) around 1200 BC, but there is evidence for iron working about the same time in Africa and as early as 1800 BC in India. Iron smelting in China was relatively late and may have been imported from the Middle East or India.

Agriculture changed large parts of the globe in a very short time. While tens of thousands of years had not seen any change from the established patterns of hunter-gathering, plant and animal domestication led to a rapid increase of other macro-inventions. Those were all logically following the new needs as determined by farming.

The early, Paleolithic times had seen the first primitive animistic religions. Animism is the belief that objects and ideas, including animals, tools, and natural phenomena are expressions of living spirits. In such societies, ritual is applied to win the favor of the spirits representing sources of food, shelter, and fertility, to provide good luck at the hunt or favorable harvests. Rituals in animistic cultures are often performed by shamans or priests, originally individuals who were believed to have spiritual powers not possessed by the other members of the band or tribe.

Initially, religion in the early Neolithic villages was not much different from Paleolithic times. Naturally, religious belief systems centered upon the weather and the seasons, as well as animal and vegetational spirits.

The worship of the moon, which was founded on the belief that its phases were related to the growth and decline of plant, animal, and human life, was widespread in the Middle East, China and India and must have obtained part of its power from its use in predicting the seasons.

In view of the close link of religion with the means of existence priests may have been the first community leaders. Neolithic village communities differed from late Paleolithic bands and tribes in their need to plan ahead. It was now important to have a calendar for telling them when to begin digging the land and planting seeds. Measuring time was not easy and only realized with a reasonable level of perfection once the seasonal movements of the moon, the sun and the stars could be correctly interpreted. In addition, Neolithic people needed to make calculations as to how much grain could be consumed to make supplies last until the next harvest, how much surplus needed to be reserved for periods of drought, how much seed was needed and how much land to dig. In these old days, measuring dimensions of plots of land must have provided the basis for the much more complicated measurements at a later time when agricultural practices were characterized by large-scale irrigation and societies by grandiose architecture. Ultimately, this would lead to science.

Hence, the societies that were now emerging consisted of sedentary people living in villages. Those were peaceful societies since no single household could initially produce the large surpluses that might have become a source of conflict. However, because of its success in creating assured food supplies more people could live on the same area of land than was the case with hunter-gatherers. Agriculture spread rapidly and it would soon meet a set of conditions optimal for creating the large surpluses that forced yet another major alteration in societal life.

Cities, states and empires

Neolithic village life naturally morphed into city life in those places where conditions were good, having led to larger surpluses. In China, stamped-earth fortified walls came to be built around settlements over the course of the 5th to 3rd millennia BC. In the Middle East, the city of Jericho may go back to as early as 8000 BC, whether as a true city or merely a settlement. Cities were a natural consequence of ever increasing food production, which was driven by population pressure, which in turn led to even larger populations. The reason that this happened first on a large scale in what is now called Iraq, is the great attraction of the Tigris and Euphrates river valleys to the hillside people. In principle, the rivers provided an unlimited amount of water for their fields. However, how to take advantage of that was not immediately obvious. In spite of their enormous fertility, due to the silt deposited by the rivers, the floodplains were not very useful for early agriculture because of flooding in winter and long periods of hot weather in summer. This necessitated the digging of canals and the building of dikes, ultimately culminating in the sophisticated drainage and irrigation systems of the Sumerians, who were the first people to practice irrigation agriculture on a very large scale. The large numbers of people involved in such irrigation agriculture led to cities that were generally larger and sometimes much larger than those that were simply the product of more modest agricultural improvements. Called Mesopotamia (land between rivers), this part of the world is often considered as the cradle of civilization. Other civilizations would follow suit and also blossom along the great rivers such as the Nile, Yellow River and the Indus.

Prehistory formally ends for a given society with the development of written records. For the Middle East this was about 3500 BC, when the Sumerians created the earliest writing, termed cuneiform, which was stamped on wet clay by a reed stylus. Different scripts were developed shortly thereafter in Egypt and later in China and Greece. Writing was a key instrument for statecraft and quickly became a much needed tool as a record-keeping vehicle for administration and commercial transactions. This allowed the development of states and, eventually, empires.

We now see the emergence of the first heads of state, probably priests. Originally serving as a conduit between the villagers and the deities, thereby taking away at least some of the uncertainties plaguing early agriculturists facing the whims of nature, the first priests must also have exercised some social leadership. Larger population centers developed as a result of agricultural success. Sumerian city communities consisted of perhaps several tens of thousands of people and had complex systems of irrigation, canals and dikes.

Cities proved to be great innovative environments, as they still are. This is due to the interconnected nature of innovation. It is a myth that great inventors conceived their ideas in isolation and owe nothing to others. In reality, virtually all macro-inventions are the products of a great many ideas that are circulating in the immediate surroundings of who is later considered the inventor(s). Innovation thrives when new ideas can serendipitously connect with other ideas.

States are sovereign political entities, which arose when leaders began to unite a number of cities and/or villages under a single government, perhaps first in the Middle East, in the third millennium, somewhat later in India, China and Africa, and later still in Europe and the American continent. While the origin of states is still incompletely understood they are likely a consequence of certain ecological conditions circumscribing a particular area where people could live, such as the Tigris-Euphrates, Nile and the valley of the Indus. War among such villages would then lead to subordination of the vanquished because there was nowhere they could flee[129].

In most of the early centers of civilization, states were eventually united into empires. Empires differ from states in the sense that they are not single political entities of fairly homogenous people. Instead, empires comprise a set of regions locally ruled by governors, viceroys or client kings. Hence, they were truly multi-ethnic states, albeit ruled from a single center. The first example of a true empire is the Persian Empire, which emerged around 500 BC. Then we see the Indian Maurya Empire around 300 BC, the Chinese Empire around 200 BC, the Roman Empire around the turn of the millennium and, much later, the Inca and Aztec empires in America in about 1400.

In the Middle East states were spreading around the Mediterranean Basin, in Anatolia (what is now modern Turkey), Greece, the Ionian islands, Cyprus, Crete, Thrace (modern Bulgaria), and Italy. This resulted in a sophisticated network of international interactions, punctuated by wars, absorptions of states into one another and, eventually, the unification of most of the Middle Eastern states into first the Persian and, finally, the Roman Empire. We see something similar in India where the Maurya Empire gobbled up virtually all independent states in what is now India and Pakistan. In China we see perhaps the most vigorously fought struggle for empire anywhere. During the so-called Warring States period, a fragmented state system was unified under the legendary Qin Shi Huang, who is considered the first emperor of China.

States and empires necessitated a much more complex division of labor and system of economic organization than in the old days of city states. Although the far majority of people continued to be active on the land as farmers or in keeping stock (as they would be until the last century), the focus

129 Carneiro R L. "A Theory of the Origin of the State". *Science*, 169, 733-738, 1970.

was now on urban life. Life was centered round such great metropolises as Babylon, Alexandria, Rome, Chang'an and later Cuzco and Tenochtitlan, which could have populations of up to a million people. Writing was a characteristic feature of the new, highly organized urban societies, and it was everywhere supplanting oral traditions and human memory. The first science emerged as a natural extension of the rudimentary forms of astronomy, arithmetic and geometry that served their Neolithic forebears so well in planning the planting of seeds, calculating how long supplies would last, measuring plots of land, etc. Applied science quickly became a cornerstone of the new societies, playing a key role in agricultural management and also serving their penchant for the grand, monumental architecture with which the age of empire is usually associated. Everybody is familiar with the pyramids of Egypt, the Great Wall of China and the glory that was Rome. The empires systematized and refined agricultural, metallurgical and architectural technology to levels simply unimaginable by their Neolithic predecessors. However, as we will see in the next chapters, in spite of their brilliant shine, great breakthrough inventions were no longer made after these empires had become stable and secure.

To summarize, technology has been an integral part of humanity immediately from its origin as a species, now almost 200,000 years ago. The difference between humans and other species, many of which also use tools, is that the human species made technology its key asset, initially to survive and later to dominate the planet. Technological progress came in waves followed by periods of stasis. The first of these waves, which led to such major inventions as the use of fire, fishing gear and missile weapons, drove the spread of humans over the entire world, leading to highly successful societies of hunter-gatherers. The great staying power of these societies is testified by the fact that they are still among us, most notably the Kalahari Bushmen and the Australian aboriginals. Unless destroyed by more powerful adversaries, hunter-gatherer societies continued their successful existence, albeit in the absence of further technological progress.

Then, in some places, changes in environmental conditions led to the emergence of the second wave of technological progress, the Neolithic Revolution, a transition from hunting and gathering to agriculture and settlement. This involved the domestication of plants and animals and occurred in different places at different times, depending on the environmental pressures and geographical suitability. It was the Neolithic Revolution that gave us virtually all the tools, albeit in rudimentary form, that we now recognize as shovels, hoes, plows, pottery, textile clothing, and dwellings. In essence, the Neolithic Revolution established humanity as we now know it. However, once established in its basic form, no further changes occurred.

The next wave of invention, which gave rise to the large population centers we are now so familiar with, may also have been driven by necessity, i.e., overpopulation, but probably also by opportunity. Like the Neolithic Revolution, it also occurred at different places and at different times. Where conditions for agriculture were exceptionally good, in sunny places with lots of water for irrigation, we see the intensification of agriculture leading to great surpluses and the development of organized social structures to manage the division of labor and the distribution of its products. Hence, this was the birth of states and governments. Their rise and competition for hegemony was associated with major inventions, such as scripts, monumental architecture, roads and cities. Eventually this would result in the rise of empires, many of which have been considered the pinnacles of human civilization. In the next two chapters we will focus on the two grandest, longest lasting and most successful empires, China and the Mediterranean Empire of Rome, and look at their rise and decline through the lens of technological progress.

Chapter 8. The Middle Kingdom: Rise and Decline of a Technology Powerhouse

The anonymous sellers of the vase found in the attic of their family home, which they were clearing out after their parents had died, must have been elated when they got the news that their treasure had become the most expensive Chinese work of art ever to sell at auction anywhere in the world[130]. The result of the auction, held on October 14, 2010, was just short of $70 million! The vase was made in the 1700s, during the reign of the Qianlong emperor of China, probably in that country's most famous center for the production of ceramics, the city of Jingdezhen in the South. This was the time of China's last imperial dynasty with the empire as large as it would ever be and at the heights of its power.

Beautiful as it may be, the symbolic value of the vase is greater still. Having invented porcelain during the Tang dynasty around 700, the Chinese held a virtual monopoly on its production until the 18th century. The techniques for combining the proper ingredients and firing the mixture at extremely high temperatures were unknown anywhere else and the beauty and sophistication of their ceramics reached such levels that in the English language "china" became a synonym for porcelain. Enormous quantities of ceramics were exported to Europe, often specifically designed for their cus-

130 Burns JF. "Qing Dynasty Relic Yields Record Price at Auction". *The New York Times*, November 12, 2010.

tomers. At the time the $70 million vase was made, porcelain production in China resembled a highly specialized, mass-production-style industry. This was almost a century before Europe began to establish a burgeoning textile industry in England in the 1750s, which is generally considered a key contributor to the complex interaction of various socioeconomic developments that led to the industrial revolution. Why did not the porcelain industry, which was at least as extensive, do the same for China? In other words, why did the industrial revolution first occur in Europe and not in China?

In China, the centralized state system began later than in the Middle East or India. Out of the diverse Neolithic cultures in the north, a Bronze Age civilization emerged between 2000 and 1600 BC along the Yellow River, where palace-like buildings and tombs have been excavated. This civilization is traditionally represented by the Xia dynasty (also called Yu dynasty, after its legendary founder Yu the Great) and considered as the traditional beginning of China's history. Shortly thereafter, with the Shang dynasty, we see the first true authoritarian state emerge in China. Like their cousins in India and the Middle East, they used writing for maintaining an elaborate organization of their state, including the cultivation of crops (mainly millet) and architectural planning. A king was the religious and political head of society.

After the Shang, China fragmented around 500 BC into a number of 'warring states'. It has been pointed out that at this point in time China was very similar to Europe in the early modern period. Like Europe more than 2,000 years later, the system of sovereign territorial states in China saw chronic warfare with its formation of alliances, the development of centralized bureaucracies, the emergence of citizenship rights and expansion of international trade[131]. I would like to extend the comparison to the Mediterranean Basin around the same time, although Chinese accomplishments were more significant. In both Warring States China and the Mediterranean Basin, and much later in Europe, the period of instability and state competition gave rise to a burst of new technology.

In China, the rulers of the domination-seeking states sought to maximize resources and productivity of their realm. They implemented a general registration of all households, and military conscription allowed them to build very large standing armies of mainly infantry. The symbolic remains

131 Tin-bor Hui V. *War and State Formation in Ancient China and Early Modern Europe.* Cambridge, New York, 2005.

of the army of the Qin were discovered in 1974 in the form of a huge army of life-size terracotta statues of soldiers near the tomb of Qin Shi Huang, the first emperor (see below); the soldiers were supposed to protect the dead emperor. (In Europe, military conscription was only introduced after the French revolution in 1792, enabling Napoleon's domination of Europe.) To feed this mass of soldiers, great increases in agricultural productivity were achieved. This was accomplished by opening up forests and wastelands, multiple cropping, seed improvements and large irrigation works. The hereditary, feudalistic society was replaced by a meritocratric system of tax-paying citizens. Not surprisingly, military technology saw rapid improvements, including the introduction of the crossbow and steel sword. Large warships were made. The multistate system also stimulated trade and the monetization of the economy.

The rulers greatly promoted the use of iron implements, especially in agriculture. In China, iron-smelting techniques were introduced only by the eighth century BC, significantly later than in the Middle East or India; they were probably brought by nomadic peoples from Central Asia. Nevertheless, the Chinese would master iron's intricacies much earlier than anyone else on the globe.

Tools and weapons were made from wrought iron, obtained by heating a mass of iron ore in a furnace using charcoal as fuel. By hammering this hot mass, impurities are removed and the now pure mass of iron can be welded in the desired form. Such a furnace was called a bloomery. Until the 14th century this was the state of iron technology in Europe and everywhere else. Everywhere, that is, except in China. China invented the blast furnace in the 5th century BC. In a blast furnace, iron ore is melted under very high temperatures achieved by blasting air into the bottom of the furnace. Somewhat later the Chinese applied coke to fuel the furnace rather than charcoal and they used waterwheels to power piston-bellows for blasting the air. The liquid iron that can then be collected is called pig iron; it contains significant amounts of carbon, which makes it brittle. The Chinese developed procedures to re-melt pig iron and, by blowing oxygen into the mixture, get rid of the impurities and part of the carbon. Depending on the desired application they could then make steel (intermediate carbon) and wrought iron (low carbon) implements. This is essentially the modern iron and steel work as used in Europe for the first time in the 19th century by Henry Bessemer (see Chapter 11).

The blast furnace allowed the Chinese to mass-produce iron and steel implements much earlier than other societies. The most important of these was the iron plow, developed in China as early as the third century BC, based on their superior cast iron technology. The Chinese iron plow had a moldboard to turn over the sod to form furrows. In comparison, the Romans continued to use the scratch plow and an iron plow of similar quality as in China at that time was only introduced in Europe in the 17th century (from China), contributing greatly to the dramatic increase in agricultural productivity in Holland and England (see Chapter 11). Chinese technology was superior to what could be found anywhere else in the world at that time and this situation would not change until the late European Middle Ages.

Paper, made from wooden fibers, was another early invention in China, possibly as early as 200 BC. As we have seen, the Sumerians had clay tablets as early as 4000 BC, to record their thoughts. A thousand years later the Egyptians fabricated sheets of papyrus from strips obtained by peeling and slicing the plant of that name. Papyrus was also used by the Greeks and Romans, but it was fragile. The Romans later switched to vellum or parchments, made from animal skin. This was much stronger but very expensive. Paper making reached Europe only in the 12th century, well in time for the printing press which could not use either papyrus or vellum.

There were many other important inventions made in China at this early time. One of the most important was the horse collar. The classical way to let a horse pull carts and plows was through a throat-girth harness, which essentially consists of straps across the neck and chest of the animal. This greatly reduced the pulling power of the horse because it restricted its breathing. During the Warring States period, China first developed the breast harness, which better distributes the load around a horse's neck and shoulders when pulling a cart or plow. This was introduced in Europe only in the 8th century. During the 5th century, China developed the horse collar as the final solution to the problem of gaining maximal pulling power from a horse by avoiding constraints on its breathing. The horse collar, which increases pulling power about three-fold, became widespread in Europe only in the 12th century. In a time when animal power played such a critical role in daily life, it is difficult to overestimate the importance of this invention, which of course is still used nowadays for draft horses. With the horse collar, land transportation was as good as it would be until the 16th century, when a system of springs was developed in Hungary such that the body of a cart was separated from the axles and was hung in leather straps.

Another macro-invention was the wheelbarrow. Although the invention of the wheelbarrow is sometimes ascribed to ancient Greece (around 400 BC) it was really first used extensively in China as early as 100 BC. Called 'wooden ox', the wheelbarrow became a fixture of Chinese life around AD 200. The wheelbarrow essentially doubled human capacity for moving goods, not unimportant in a time when a lot relied on human or animal power. The first convincing evidence for extensive usage outside China was in Medieval Europe around AD 1200.

The Warring States period also saw the beginning of the 'Grand Canal', a monumental engineering feat to connect China's many rivers creating a transportation network that would eventually be more than 1,100 miles long, with many locks and bridges. An important consideration in its initial construction was probably troop transport. Another monumental construction began at this time was the Great Wall, a series of stone and earthen fortifications to protect the northern borders.

Finally, the Warring States period also witnessed the emergence of the great thinkers, such as Confucius and Laozi (5th century BC), Mengzi (372–289 BC) and Xunzi (298–238 BC). Similar developments took place at about the same time in India and Greece. Hence, it was during the Warring States period that technology development in China took off. It was then when China abandoned slavery and feudalism and became a relatively open society with a free peasantry. Although later under the Han and Tang dynasties we again see the emergence of rich nobles in China, they would never inherit their positions as state functionaries or rights to rule their land as independent potentates.

The struggle for dominance during the Warring States period has been compared to the chronic conflicts that characterized Europe until fairly recently, which led to catastrophes as the First and Second World War. But while in Europe there was never a state that could call itself victorious, China was unified in 221 BC by Qin Shi Huang (259 BC–210 BC), the king of the State of Qin. He became the first emperor of a unified China. Shortly after his death, a revolt led to the end of the Qin and the flowering period of the Han dynasty. Under the Han, the Chinese empire gradually expanded in all directions from its original core along the Yellow River in the north (Figure 1).

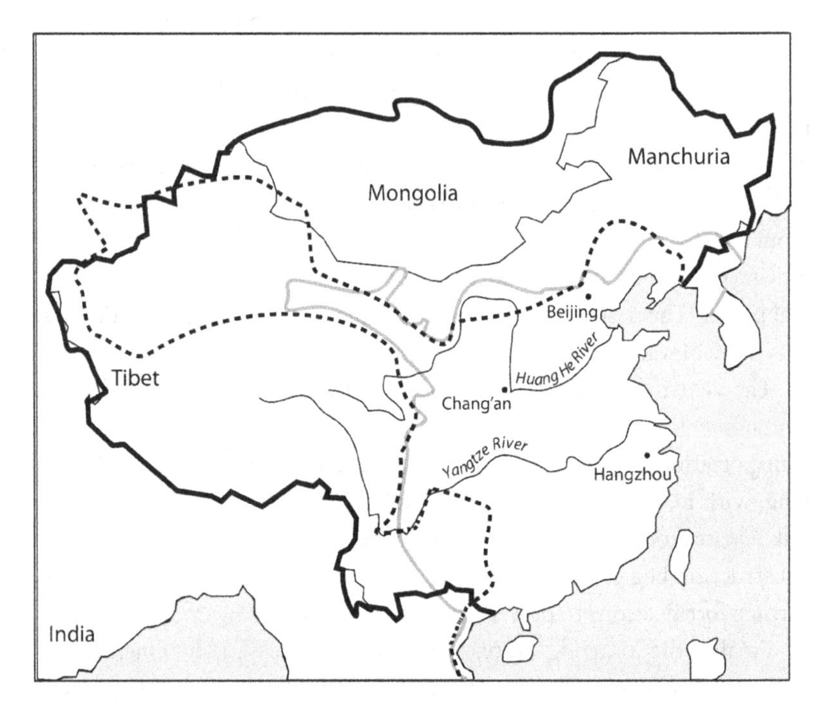

Fig. 8.1. Borders of Han, Tang and Qing China. It is obvious that the empire reached its highest extent under the Qing, i.e., around 1800. Grey lines: Han; dotted line: Tang. Solid, black line: Qing.

The Han colonized the Yangtze Valley and included oases along the Silk Road into their empire. Private landholding with a focus on small farmers was now dominant in China, in striking contrast to the situation elsewhere, e.g., the Roman Empire, early Europe). Agriculture became more and more based on intensive cultivation involving sophisticated techniques of irrigation and seed selection. More and more extensive use of the iron plow further increased productivity with more marginal regions brought under cultivation. During the Han dynasty Chinese ironworking had already achieved a scale and sophistication not reached in the West until the 19th century. During Han times, the Chinese also invented deep borehole drilling for mining and other projects, such as lifting brine to the surface through bamboo pipes for salt processing.

A major invention was the waterwheel, one of the few ways the ancients could create mechanical energy independent of human or animal power.

Waterwheels use flowing or falling water to move paddles fastened to the wheel. This results in rotation which can be converted into power via the shaft of the wheel. Waterwheels were used for crop irrigation, grinding grain and later to drive sawmills, power the piston-bellows of the blast furnace and operate trip hammers for crushing ore. Waterwheels come in three variations: the horizontal wheel and an undershot and overshot vertical wheel. They were invented first in the Middle East between the 3rd and 1st century BC, when technology development reached a high level during the Hellenistic period (see Chapter 9). However, under the Han, China invented the waterwheel independently, initially a horizontal wheel.

It was also under the Han that many of the features of the Chinese imperial system, which would last for almost two millennia, took shape. Bureaucratic government based on Confucian values, with its civil servants selected through an examination system, became the norm in China and would last until far into the 19th century.

The Han dynasty eventually succumbed (around AD 200) and China fractured again in different states before being reunified in the 6th century under the Sui and quickly thereafter the Tang dynasty. The Tang empire, established by Li Yuan, who was of Turkish descent, surpassed the Han and represented a new high point in China's already glorious history. The Tang period, which lasted until AD 907, was a very cosmopolitan time. China's power reached into Korea in the North, Indochina in the South, Kashmir in the South-West and in the West as far as modern-day Tajikistan (Figure 1). Indeed, this was the golden age of the Silk Road, with a continuous stream of merchants benefiting from the commerce between East and West. At this time, the Chinese empire welcomed foreign cultures and there lived a significant number of foreigners in its cities. The Tang also had a good handle on the danger that came from Inner Asia, which precluded complete stability in China until the future Ming and Qing dynasties. The free-roaming tribes in these steppe lands (Manchuria, Mongolia, Tibet) were professional fighters and more than a match for China's conscript infantry armies. Being from Turkish descent themselves, the Tang emperors may have had a better grip on these people and they managed to control them.

The Tang capital, Chang'an, was the world's greatest city at the time, with more than a million people when London or Paris had at most 10,000 to 20,000 inhabitants[132]. For a time without highways and motorized traf-

132 Cotterell A. *The Imperial Capitals of China*. Pimlico, London, 2007.

fic, infrastructure was magnificent. The Grand Canal of China, first begun around 500 BC, developed into a continuous waterway cutting across the Chinese mainland. It would eventually (under the Ming) connect Beijing in the north with Hangzhou in the south.

Schools came into existence to teach Confucian classics. This was greatly facilitated by the new invention of printing from engraved wooden blocks. Since the beginning of the Tang reign in the 7th century, there had been a drive southwards to settle the (drained) marshes of the coastal region and start cultivating wet rice on the coastal plains. More iron implements, such as seed drills, rakes and harrows were introduced in agriculture. Fertilizer now came in a large variety, from hemp stalks to urban refuse. Also insect and pest controls were used, on the basis of both chemicals and biological agents. China under the Tang invented the world's first clockwork escapement mechanism (in 725), as well as porcelain, and made significant advances in structural engineering and medicine. There was also no shortage of gadgets, such as a mechanical wine server that used a hydraulic pump to siphon wine out of a faucet.

Eventually, in 907, the Tang dynasty collapsed because of internal rebellions. The resulting period of fragmentation ended in 960, when most of China was reunited by Zhao Kuangyin, a general of the Later Zhou Dynasty, one of the Tang successor states. A new dynasty, the Song, was established with Kaifeng as the capital. Under the Song ancient China reached its highpoint[133]. More than ever before the Song stressed the superiority of a meritocracy. While as we have seen meritocracy and the abandonment of feudalism and slavery first emerged during the Warring States period, remnants of a class society with rich and powerful land owners were always present under the Han and Tang dynasties. The Song perfected the ideal of the scholar-official, selected through the examinations in which everyone, including the most humble citizens, could participate. This method of selection, leading to the *jinshi* degree (somewhat similar to today's doctorate), soon became accepted as the only legitimate way to reach high office. This system, open to talent and egalitarian, produced China's stars in the arts, literature and science. It was also the pool from which the emperors drew their advisers and officials. Focused on analytical thinking and originality, members of the scholar elite were not hesitant to display an open mind and give their honest opinion about every aspect of society including matters of

133 Mote FW. *Imperial China 900 – 1800*. Harvard University Press, Cambridge, 1999.

state. Their Confucian-minded advice to the ruler was to be benevolent and righteous. China's scholar-officials were now guiding the holder of power toward socially responsible governing. Under such a system, the military were strictly subject to civilian control, and the Song carefully refrained from leaving their generals in command over the border troops.

By the end of the 12th century, perhaps as many as a hundred million people in China were governed by about 40,000 state officials. This made it by far the most populous and most sophisticated state on earth. They had an efficient tax collecting system, tapping the enormous wealth that was a product of a remarkable economic growth, a consequence of further technology development, and an elaborate political system to maintain the peace with their barbarian neighbor states. It was again further improvements in agriculture that drove all other developments. New strains of rice were introduced, including early-ripening rice from Vietnam, which allowed two or more crops a year from the same parcel of land. Along with improved methods of water control and irrigation, this increased food production dramatically, doubling the population.

Great iron works produced as much iron and steel as the rest of the world together. Instead of charcoal, these were now fueled by coke, obtained as a residue of heated coal to drive off impurities (similar to making charcoal from wood). Coke would be used in Europe only in the 19th century. The Song further extended the Grand Canal, bringing its length to 1,114 mi (1,795 km), which made it the longest navigable canal in the world, which it still is. They further improved transportation technology by inventing in AD 984 the double-gate lock, which let gravity shift ships from one level to the next, instead of haul the boats up inclined planes. England only began to build such locks during the 18th century during the industrial revolution. As in China, but more than 600 years later, canals were also the basis of industrialization in Europe and the United States.

Apart from the completion of an integrated system of internal waterways, the Song also developed new, improved ships. They invented paddlewheel boats, another variation on the waterwheel. As we have seen, the waterwheel was used for moving irrigation water and providing mechanical energy. They now fastened paddle wheels on a ship to move it forward. This was done by transmitting the power generated by humans or animals through a gear train to the paddles, which turned the wheel through the water and moved the ship forward. Far later, in the late 1700s, paddlewheels were combined with a mechanical power source (the steam engine) in the United States and Europe.

Numerous technical advances were made to the famous junk, originally developed during the Han dynasty, such as their sail plan, hull design (hulls were compartmentalized, which greatly reduced their chance of sinking) and stern-mounted rudders, which were later adopted in Western shipbuilding. These improved vessels and the invention of the magnetic compass permitted oceanic travel. In addition to the coin system, credit instruments were developed. All together this greatly increased trade, with now 10% of the population living in cities.

The Song also saw the birth of gunpowder. It was used to create bombs that were catapulted into cities. The use of gunpowder in cannon was less popular in China, possibly because their city walls from rammed earth could not easily be breached by a cannon ball. As we will see later, the European brick or stone walls around castles or cities made cannon highly useful and helped to end the reign of the petty lords. Cannon were nevertheless used in China in the early 14th century.

As mentioned, printing may have begun under the Tang (as early as AD 800), but under the Song movable type was invented, first using wooden pieces and later ceramic and metal types. This was 400 years before Gutenberg developed a similar device in Europe. Books became widely available, which greatly increased literacy. But it was really Korea that was at the forefront of printing. While woodblock printing may have been invented under the Tang in China, the oldest example of woodblock printing is of Korean origin: the *Pure Light Dharani Sutra*, printed around 750. When I was in Gyeongju, the ancient capital of the Silla Kingdom, I was told about its discovery in the temple in front of me, the Seokgatap Pagoda. It can now be admired in a museum in Seoul. Allied with Tang China, Silla was one of East Asia's most sophisticated states.

Korea was also a pioneer in movable type, which was more advanced in that country than in China. In Korea, the transition from wood type to metal type occurred in the 13th century, to meet the high demand for religious and secular books. Koreans also resolved the Chinese dilemma of the unwieldy number of characters by devising a simplified alphabet of 24 characters called Hangul, in 1443 by King Sejong, for use by the common people who were less proficient with the Chinese script used at the time. It is still in use today[134].

During the Song Dynasty accurate water clocks were built using escapement mechanisms. An escapement regulates the energy the clock receives by portioning it out into small regular bits of movement through the teeth of a gear. In China, the principle was first used by a Buddhist monk as early as AD 725. The highlight of these first mechanical clocks was undoubtedly the one designed and built by Su Sung, as ordered by the emperor, which was completed in 1094. Time in this clock was kept by a large waterwheel which was built in a 40-foot tower. The water-driven mechanism also reproduced the movements of the sun, moon and several stars. However, the first practical household clocks were invented by Europeans in the 14th century. The Chinese may generally have been satisfied with the public time signals[135].

Hence, China under the Song developed into a brilliant, refined society with sophisticated technology, driving wealth to great heights by applications in agriculture, industry and commerce. At this point, China reached its apogee as the richest and culturally most advanced society on Earth. It had the world's greatest commercial networks, contained the world's largest cities and printed large numbers of books in a time when the rest of the world had hardly introduced paper (from China). Under the Song, China was an industrial state. Like India later, it mass produced the products that people wanted, which were exported on a large scale. Europe would only surpass this in the 19th century. Indeed, under the Song, an estimated one third of all manufactures in the world were produced by China.

Yet, in contrast to the British Empire during its 19th century heydays, when through its battleships and machine guns it created absolute military superiority, the Song had to maintain a precarious military balance with two powerful neighboring states, the Khitans in the north and the Xixia in the North West. While these states were merely tribal societies of nomads and no match for China's flowering civilization, from a military perspective they were superior. This was entirely due to the dominance of the horse in warfare until deep in the 19th century. Cavalry alone could provide mobility, which multiplied the fighting power of even the smallest forces by giving them the advantage of outflanking their enemy or retreat and escape. China

134 Cumings B. *Korea's Place in the Sun*. Norton, New York, 2005.
135 Landes DS. *Revolution in Time*, Harvard University Press, Cambridge,1983.

simply did not have as many horses or trained cavalry as their steppe enemies. As we will see in the next chapter, the rulers of the Roman Empire experienced similar problems, but were always more adept in responding to it.

In 1127, one third of China was lost to the Jurchens, a tribe that destroyed the Khitan empire to subsequently establish its own Jin empire. Even under these circumstances the Song (which became the Southern Song) continued to flourish. It took until 1271 for the Song Dynasty to disappear under the relentless assaults of the Mongols, who finally took over the entire country to establish their own dynasty, the Yuan Dynasty. It was the founder of this dynasty, Kublai Khan (a grandson of the great Genghis Khan) who was visited in 1275 by Marco Polo, the Venetian who expressed amazement in his published travel accounts about China's enormous power, great wealth, and complex social structure[136]. Under the Yuan, the Chinese capital became Beijing at the northern end of the Grand Canal.

Although the Yuan Dynasty was eventually accepted by the Chinese scholar elite as legitimate, their policies favoring Mongols and other Inner Asians were resented. This led to uprisings resulting in China's fragmentation into multiple rebel states. The ruler of one of these states finally managed to gain the upper hand and reunite China (in 1368). The new dynasty, called the Ming (Ming means 'brightness'), would be the last ruling dynasty of China that was ethnically Han Chinese. With the Ming began one of the greatest eras of orderly government and social stability in human history. China's population, which had suffered greatly during the Jin and Mongol invasions and the resulting civil war, now began to increase again from about 70 million in 1400 to 100 million in 1650 to over 300 million at the end of the 18th century.

Life under the Ming was only briefly interrupted when in 1644 the successful invasion of the Manchus (descendants of the Jurchen living in Manchuria) took power and established the Qing Dynasty. Like the Mongols before them, the Manchus seamlessly emulated their Chinese predecessors. They retained most institutions of the Ming, adopted the Confucian civil service system and enforced the Confucian philosophy, further emphasizing the obedience of subject to ruler as the state creed. Hence, both under the Ming and Qing, domestic order was firmly maintained, which led to unprecedented peace and prosperity. This age of stability would last until

136 Latham R (translator). Marco Polo: The Travels. Penguin, New York, 1958.

the early 19th century. In 1912 the last, child emperor was forced to abdicate, leaving behind an empire of chaos with hostile states infringing on it from all sides.

NEEDHAM'S PARADOX

We have now reached the point where the question with which I started this chapter becomes most relevant: Why did the industrial revolution first occur in Europe and not in China? This question was raised by Joseph Needham in his lifetime project *Science and Civilization in China*, a monumental achievement of 20th century scholarship and still ongoing. Originally a biochemist, Needham learned Chinese and traveled extensively in China during the 1940s, visiting historical sites, meeting with Chinese scholars and collecting lots of books. Upon returning to England he began with the project, with an international team of collaborators, publishing its first volume in 1954. He personally wrote a number of volumes and parts before he died in 1995. The project is still ongoing, with at the time of writing seven volumes published by Cambridge University Press[137].

The Needham project was initially devoted to compiling a list of all inventions and ideas made and conceived in China, including the major ones I described above. The Needham question is of course how such a successful state as China, which before the 19th century seems to have made virtually every major invention first, could ever have lost its primacy to Europe[138]. While many potential explanations for this shift in the world's technological focal point from east to west have been postulated, including Needham's own somewhat vaguely formulated impact of Confucianism and Taoism, the most logical reason for China's surprising failure to capitalize on its enormous lead in technology seems to be foreign encroachment, in this case by Western nations and Japan. As we will see in the next chapter, this situation was the same for the mighty Roman Empire which was brought low by barbarian invasions. Yet, these simple explanations are generally rejected, I think mainly because of two reasons. First, it is often considered unlikely that a powerful empire simply throws in the towel when threatened by some barbarians from outside. Second, and this most definitely applies to China, in the 19th century most Western observers refused to acknowl-

137 http://www.nri.org.uk/science.html
138 Lin JY. "The Needham puzzle: why the industrial revolution did not originate in China". *Economic Development and Cultural Change*, 43, 269-292, 1995.

edge China's superiority as a stable, wealthy civilization that could have continued for centuries were it not for their own predatory practices. This attitude began to promote a way of thinking, soon generally accepted, that ascribed China's decline to internal rot due to the wrong state of mind. In other words, they had it coming!

It is argued that China's decline began under the Ming. For example, such eminent western scholars as David Landes argued that China was no longer competitive with Europe as early as the 15th century. He ascribes that to an absence of incentive for learning and self-improvement[139]. After all, in contrast to Europe, China was unified under one emperor; opportunities for others who aspired to wealth and knowledge were limited. In Europe, with its multiple states usually at war with each other, a certain flexibility existed. Very similar to China during the Warring States period, there were always some places in Europe where new ideas were readily adopted because they could increase the power of the state. And states were competing for talent, offering opportunities to discuss such new ideas and bring them into practice.

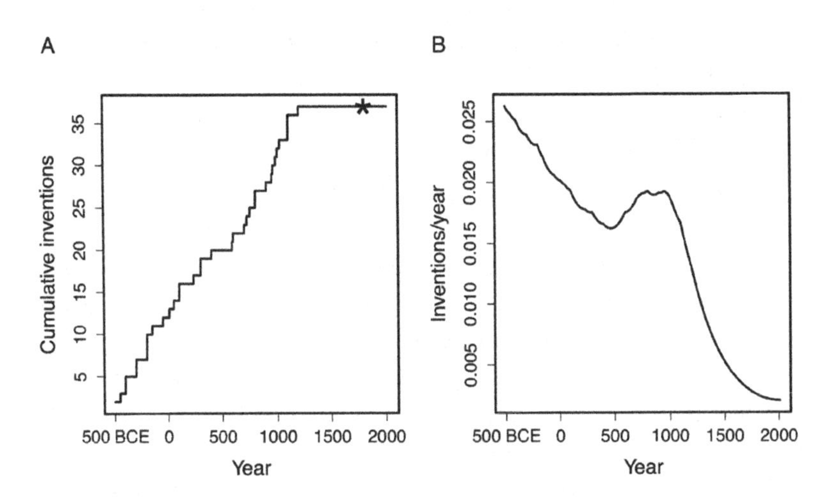

Fig. 8.2. (A) Cumulative number of major inventions in China since 500 BC. Asterisk indicates the year 1800. (B) Chinese inventions per year over the same time period. Inventions were taken from the list of macro-inventions with exception of the inventions that were independently made in China, but later than elsewhere.

139 Landes D.S. *The Wealth and Poverty of Nations: Why Some Are So Rich and Some So Poor*, Norton, New York, 1998.

And, indeed, around the 15th century China's inventive streak had definitely come to a halt. The steady stream of innovations that had become so characteristic for China since the Warring States period was no more. This is clear from Figure 2, in which the accumulation since 500 BC of all macro-inventions first made in China are plotted (compare with Figure 4B in Chapter 1). From this graph it is obvious that new inventions were accumulating rapidly during the turmoil of the Warring States period (476–221 BC), and also later until at least the 12th century. After that, however, not a single breakthrough invention can be reported. This seems to confirm the general assumption that already at the time of the first contact between China and the Western powers (at that early time, mainly Portugal) Chinese society was inferior. Nothing, however, could be further from the truth.

In fact, when we consider contemporary opinions rather than current scholars having the wisdom of hindsight, there is no evidence of a Chinese decline in the 15th century at all. While it is true that in the 15th century a surging Europe most definitely began to close the technology gap, China remained far ahead of the rest of the world as its most successful state. Indeed, it was only during the Qing Dynasty late in the 18th century that the empire reached the height of its power. While no further major technological breakthroughs were made, there simply cannot be a misunderstanding about the success of Chinese society in providing a safe, stable environment for its people to thrive. The argument that already in the 15th century Europe had the upper hand in economic power and weaponry is simply wrong. In any head to head confrontation, Europe would have been crushed.

Chinese society since the Warring States period had developed into a relatively open society, with land privately owned, a market economy, a high degree of division of labor, a magnificent transportation infrastructure and a level of technological development that would remain unmatched until late in European history. Under the Ming, China established sea routes for trade with Japan and south Asia. In 1405 the famous Admiral Zheng He began his series of naval expeditions that went as far as the east coast of Africa. These trips followed established routes and were mainly diplomatic. The last of these voyages was completed in AD 1433. At this point, China was so far ahead of the rest of the world, not only in naval capabilities but also more generally in economic and technological development, that a Chinese rather than a European colonial empire was a more likely prospect. However, the

Chinese lacked an incentive for doing that. Unlike the Europeans, China already had everything, and that would remain so for a long time.

Indeed, economically China remained far ahead at least until the 18th century. Agricultural technology continued to contribute to productivity increases, undoubtedly because private property rights over their land created incentives for the small farmers. Also the total acreage of cultivated land reached new heights. As we have seen, the population surged from about 70 million in 1400 to over 300 million in 1800, mainly because of further innovations in agriculture, including the adoption of maize and sweet potatoes from America.

Urbanization increased as the population grew and as the division of labor became more complex. Both large and small urban centers contributed to the growth of private industry, specializing in paper, silk, cotton, and porcelain goods. In those days China produced over 100,000 tons of iron each year, which remained more than the rest of the world combined. More books than ever came off the printing presses and traditional scholarship and arts flourished. Even in rural areas schools were common and basic literacy relatively high. Slavery was forbidden and independent peasant landholders predominated in Chinese agriculture. A genuine money economy had now emerged with large-scale mercantile and industrial enterprises, many under private ownership, most notably some great textile centers in the southeast. China's borders also expanded. Manchuria, Mongolia, Xinjiang and Tibet were all brought securely under Qing control, making the empire larger than it had ever been in the past (Figure 1). For the first time in 2,000 years, the northern steppe was not a serious threat to China's defenses.

Therefore, in spite of its technology slowdown, China remained healthy and the economic superpower of its times. However, Landes and others are certainly correct that with the Ming and Qing dynasties, a strong despotic streak was introduced into the government, which effectively constrained the individual, Confucian sense of responsibility that had been so characteristic of China in previous centuries. A centralized administration was a fact of life in China at least since the Warring States, but life was now so dominated by a strong bureaucracy resisting change at all levels that few incentives remained to make breakthrough inventions. Many of the most talented individuals chose to stay out of government, trade or other business and instead focus on art and entertainment. Many a great novel was written in Ming times, and Ming painting and pottery still rake in record prices. Similar to what we see in our current 21st century society, the increased stability and well-being in China under the Ming and Qing led to a shift to less productive areas in society and began to dampen innovativeness.

While not developing itself as the first industrial state, there is no evidence that China suffered from major internal decline. Most notably, eighteenth century Europeans did not seem to have shared the opinion of Landes and other current historians that China was on the decline as early as the 15th century. The Italian-born Jesuit priest, Matteo Ricci, who was sent as an emissary to China in 1582 and died there early 17th century, declared that he is "filled with admiration for the great Chinese Empire'[140]. As mentioned by Frederick Mote (1922–2005), the American sinologist and professor of history at Princeton University, in his book *Imperial China 900–1800*[141], "Westerners who became familiar with conditions in China in Qing times often compared China favorably with Europe, remarking that the masses of ordinary people were well ordered, cheerful, and mannerly, mostly well-fed and well-housed, and with great capacity for energetic pursuit of their personal and family interests."

And while the Qing emperors were interested in the West, mimicked some European art and collected clocks, watches and telescopes, they did not feel there was anything to be gained by close interactions. This was clear from the letter by Emperor Qianlong to King George of England after the Macartney Mission, a British embassy to China in 1793. Lord George Macartney was sent to convince the Emperor to ease restrictions on trade between Great Britain and China by allowing Great Britain to have a permanent embassy in Beijing. But the Emperor wanted none of it. In his letter, Qianlong explained the reasons for his refusal to grant the British requests, pointing out that China was not interested in and did not need foreign items.

From this we see that, on the eve of the 19th century, the emperor of China still saw no benefit being in permanent contact with Europeans. And of course, as suggested by the above quote from Mote's book, it is highly questionable whether the average European was better off than the average Chinese in those days. China under the late Ming and Qing dynasties

140 Laven M. *Mission to China: Matteo Ricci and the Jesuit Encounter with the East*. Faber and Faber, London, 2011.

141 Mote FW. *Imperial China 900–1800*, p. 941. Harvard University Press, Cambridge, 1999.

until the early 19th century was a stable and secure society where people most of the time lived uneventful lives. There was food security, a high level of hygiene and a robust health care system that could deal with a host of diseases, including infectious disease such as typhus and smallpox (inoculation against smallpox was widely practiced since the 16th century). While unlike physicians in early modern Europe, such as Vesalius (1514–1564), Chinese physicians did not practice dissection or any direct anatomical studies, health care in ancient China was quite advanced with an extensive pharmacopoeia.

Chinese real wages were probably higher than in Europe at least until the 18th century. It is true that at this point in time Europe had passed China as the most innovative place on earth and it quickly became technologically superior. However, for China to acquire many of the accomplishments of Europe was very similar to our own current option of having supersonic airplanes, smart highways, moon bases. Do we need them? Many would still say we do not. So did the Chinese in 1793. The consequences for them of course were similar as the consequences for us: stalling wealth generation. But for the Chinese Empire, like for the Roman Empire before them, there were other consequences as well. They would always be subject to aggression from outside.

In fact, it was really only after the Opium Wars (1839 to 1842 and 1856 to 1860) that the process began that eventually led to the collapse of the empire. It was the import of opium by the British East India Company that turned the empire's stupendous trade surplus with the West into a deficit. The subsequent Chinese attempt to prevent imports of opium into Canton led to the first embarrassing demonstration of its technological inferiority during the subsequent opium wars. While China was at a standstill, the European nations had continued their technological revolution that began around the same time China's inventiveness had come to a halt. Even in our present time, China's wealth and power vis-à-vis Western Europe is far from being restored to the heights it reached at the time of the Qing Dynasty.

In summary, up until the early 19th century, contemporary observers, including European observers, would not have understood the opinion that China under the Ming was experiencing the beginning of a slow decline into oblivion. They admired and respected what was obviously the greatest state on earth. For us it is easy to argue that China was doomed to go under because they had become averse to change. We tend not to like the increased despotism that began to dominate Chinese society under the Ming and Qing, from about AD 1400 onwards. And we therefore automatically assume that their people were in bad shape, an assumption that has been greatly supported by the Chinese communist party once they were in charge of the state. Indeed, the disasters associated with the reign of Mao have made it all too easy for the communist ideologues to make their people believe it was all much worse under the emperors. The reality, as modern scholarship has made clear, is that 18th century China was a much better place to live than most places anywhere else in the same period or, for that matter, China in the 1950s.

Let us now recapitulate what we learned from the Chinese experience with respect to technology and innovation. First, there is not much you can do without optimal natural and geographical conditions. As Jared Diamond explained so convincingly, agricultural powerhouses do not take off without naturally occurring plants and animals that can be domesticated or fertile soils that can easily be worked. Like everywhere else agriculture was the key implement in the rise of sedentary human civilization, whereby farming of domesticated species created food surpluses that nurtured the development of much denser and more stratified societies.

But even if the environment is kind, development into a successful society does not come automatically. The other major factor that seemed to have potentiated China's rich technology palette, especially its precocious success in generating advanced farming and iron and steel works, appears to be vigorous interstate competition. It was really during the Warring States period when China pulled itself away from every other civilization, including the Mediterranean society that developed around the same time. Somewhat surprisingly, China's history also shows that even after the source of innovation has dried up, a stable, successful society is able to maintain prosperity and a high level of technological prowess for centuries without substantial new inventions.

If conflict promotes breakthrough inventions and vigorous innovation, then, let us now see if we can find a parallel of China's history in the other great empire that continues to appeal to our imagination, the Roman Empire.

Chapter 9. Rise and Fall: Rome as the Great Synthesizer

Around 517 BC, the biblical prophet Daniel saw four great Empires that would rise and rule the world. This vision was presented as a huge statue of a man made of four different metals corresponding to the four empires. The fourth empire, Rome, was represented by the legs and toes of iron and clay and considered a military state that lived by the sword. The feet of clay symbolized its eventual collapse and the question of why Rome went under has grown into one of the most discussed topics among historians. Here we will look at it from a technological point of view.

Often compared and overlapping in time the Roman and Chinese empires could not have been more different. The Chinese empire since its founding by the Qin was regularly fragmented and re-unified, often by invading northern steppe people, the Roman Empire was a stable entity from at least 202 BC (when the victory over Carthage opened up the Mediterranean basin to Roman supremacy) to AD 1204 (when a crusade of Western Europe invaded and conquered the city of Constantinople, capital of the Roman (then Byzantine) Empire. While formally the Empire was briefly restored until 1453, it never managed to regain its former strength.

Similar to the Chinese Empire, the Roman Empire emerged from a group of brilliant, highly competitive states. In this case, the competing states were grouped around the Mediterranean Basin, with a heavy focus on the Middle East (Figure 1). Like China, they were a dazzling showcase of technological acumen. Some of these states were linear descendants of the Sumerian and Egyptian civilizations, which were absorbed in the 6th century BC by the Persian Empire, often considered as the first real empire and also the geographically largest of them all. This super state also managed to include the Greek cities of Asia Minor (now Turkey), but the tables were turned when Alexander the Great of Macedonia conquered not only Greece but also the entire Persian Empire in 331 BC (hence, the 'Great'). With the division of his empire among his generals after his death less than 10 years later began the Hellenistic period, a fusion between the Greek city states and the great Middle Eastern civilizations, which would last until the first century when Rome managed to conquer them all. It was the Hellenistic period that would turn out to be the last genuine age of invention before the unification of the entire Mediterranean Basin by Rome.

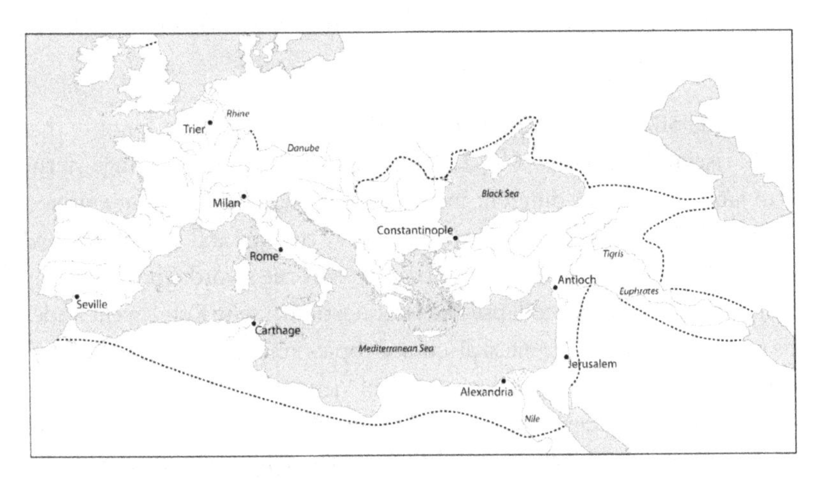

Fig. 9.1. The Roman Empire at its greatest extent, during the second century.

Like in China's Warring States period from the 5th to the 3rd century BC, at about the same time the vigorous, competitive interactions among the Mediterranean states, including not only the Hellenistic states but also the nascent Roman state, and the merchant state of the African city of Carthage, spawned innovation. Much of that went back to the Greek oikoumene, which by 300 BC had spread out from the shores of the Black Sea, the port cities of Asia Minor to South Italy. It became the cradle of abstract science (or philosophy, as it was known in those days). Like in China and India around the same time, it was in the Greek cities in Asia Minor where humans first began to address questions about the nature of the world and humankind, no longer in religious terms but in rational, objective ways. The most famous early philosopher was Thales from the city of Milete, who lived from 624 to 545 BC. He was followed by many other Greek philosophers, all eager to understand the world. Among them are Pythagoras, of legendary fame in mathematics, and eventually the great philosophers, Socrates, Plato and Aristotle, who still to a great extent dominate western thinking.

The Greeks were also the first to accumulate empirical knowledge in a variety of scientific disciplines, including human anatomy and physiology, metallurgy, mineralogy and astronomy. Aristotle's works on natural history are a good example. In his grand biological synthesis he grouped together animals with similar characters into genera. This was based on real observations of their anatomy, many of which remarkably accurate.

The Greek city states were typical examples of communities where agricultural surpluses were not high enough to lead to super cities, kingdoms and empires like we have seen in Mesopotamia and Egypt. Instead, like the Phoenicians from what is now known as Lebanon, the Greeks practiced trade on a large scale, which as we now realize is a major source of economic growth. Apart from their wives and their slaves, Greek communities were more or less democratic as we would expect in the absence of the centralized production of great amounts of food. Perhaps their trade and their democracy gave them time to think in original ways, which would then explain why such key developments in science and technology first happened there and not in one of the larger and more powerful states on their Eastern borders.

No other human being is so much associated with science and technology as Archimedes. Born in Syracuse, a major Greek city on the island of Sicily, Archimedes (287–212 BC) is often considered as the greatest mathematician of all time. He developed pi and in our time is best known for his Principle: *an object immersed in a fluid experiences a buoyant force that is equal in magnitude to the force of gravity on the displaced fluid.* He also invented useful tools, such as the hydraulic screw — a helical pump for raising water from a lower to a higher level (Archimedes screw), the catapult, the lever, the compound pulley and the burning mirror. But there were others as well. Eratosthenes (276–194 BC), the chief librarian of Ptolemy III of Egypt in Alexandria is still well-known for his 'Sieve', a logical procedure for calculating the prime numbers. He also measured the circumference of the earth (it is a myth that educated people, even in the European Middle Ages, thought that the earth was flat), calculated the distance to the moon and produced a calendar, which via Julius Caesar would become the basis for today's Gregorian calendar.

Yet, brilliant as these achievements of the Greeks and their Hellenistic successors were, these highlights of innovation would not last and with the unification of the entire Mediterranean Basin by the then still republic of Rome, the creative streak of the Mediterranean Basin was rapidly silencing. This becomes clear from Figure 2, in which I plot the new inventions originating from the Mediterranean Basin since about 500 BC. As discussed, the period from 500 BC to the beginning of our era, i.e., the year zero, witnessed a highly creative period, characterized by violent inter-state conflicts. With the end of that era centuries of prosperity and relative tranquility would begin. Later historians would call this the *Pax Romana* and the happiest period of mankind. Indeed, in the first century, the Roman Empire stood only at the beginning of its most grandiose accomplishment of unifying many disparate people geographically separated over thousands of miles and letting them live and trade in peace under the benign, but strict sovereignty of an enlightened despot. However, the Roman peace brought stability but it failed to discover new and better ways of producing goods and agricultural products.

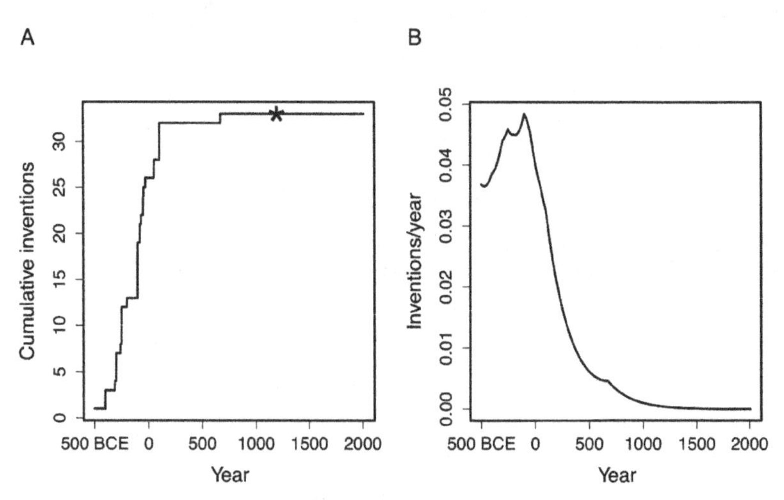

Fig. 9.2. (A) Cumulative number of macro-inventions in the lands of the Roman Empire since 500 B.C. Asterisk indicates the year 1200. (B) Inventions per year. Inventions were taken from the list of macro-inventions with exception of the inventions that were independently made, but later than elsewhere.

Hence, similar to the situation in China we see a rapid increase of inventions during the period when fierce competition among states was endemic, but once a strong, stable state had ended all quarrels and established a wealthy, prosperous and peaceful society this all came to a halt and inventiveness dried up. This technology slowdown happened significantly sooner and was more abrupt than in China where as we have seen innovation was still high under the Han, Tang and Song dynasties, which were also stable times. However, we have to realize that most technology in the Roman world was developed not by Romans but by the Hellenistic kingdoms and their predecessors in the Middle East. Also, China was never as stable as the Roman Empire and underwent repeated bouts of fragmentation. In China nomads of inner-Asia were always a threat that the emperors found difficult to counter. We will see later in this chapter that while this was true for the Romans as well, they were much more capable to counter barbarian invasions than the Chinese.

Roman society also differed from the Chinese in many other ways. Perhaps the most important difference was that in contrast to China, Roman society as all Mediterranean states that went ahead of them practiced slavery on a large scale. While farming in Rome was originally practiced by free peasants, already at the end of the republic, during the first century BC, most agriculture took place on great landed estates called latifundia, which depended upon slave labor. Slaves were also working in the mines, as domestic servants and as teachers. There were of course areas where slavery was less important and there were also free peasants, but on average Roman economy was not based on free citizens.

Slavery was an inherent part of the Roman state, which in contrast to China was originally a robber state with the empire a product of aggression and domination. Rome's golden age was based on chronic warfare providing the rulers with captives who were sold as slaves. When Augustus, the adoptive son of Julius Caesar, put together the first responsible administration worthy of that name, seasonal wars and the opportunity for plunder greatly diminished[142]. On top of that, Roman society saw a rise in humanitarianism (some authors from those days called that 'increased feminization'), which led to frequent manumission. While slavery was never abolished, rules were

142 Everitt A. *Augustus: The Life of Rome's First Emperor*. Random House, New York, 2006.

instituted to protect slaves against the worst abuses. Hence, increased success in creating a stable and wealthy empire also meant the end of cheap slaves.

To some extent, slavery could explain why the Romans have generally been considered as technologically uninventive. After all, an abundance of cheap slaves is hardly conducive for setting in motion large scale efforts to improve productivity. But that does not explain why the quality of almost all they produced lagged behind its Chinese equivalent. In the previous chapter, I discussed their lack of insight as to how to enable a horse to pull a cart or plow more effectively. They did not have paper, and compared to the Chinese, they used primitive agriculture and iron works. While iron was occasionally used in innovative ways (for example, to reinforce their concrete vaults), the Romans did not produce iron or steel in bulk, like the Chinese (and later the Europeans), who used bellows and hammers driven by waterwheels. The explanation for the Roman lack of innovation is more likely to involve a lack of incentives rather than a sheer abundance of unfree human labor. Slaves have no incentives to invent or innovate and the same was true for the many semi-free tenants working for the big landowners. This was certainly different from the situation in China, where a free peasantry was the bedrock of society.

Like China under the Ming and Qing dynasties, the Roman Empire experienced a technology slowdown but not a technological decline. They maintained the technological tools they inherited from their Hellenistic forebears. Moreover, they excelled in architecture and engineering, areas that greatly benefit from inexhaustible manpower. This explains why a lack of technology is not a thought that crosses one's mind when confronted with the remains of the Roman Empire. Roman engineering, responsible for such stupendous works as the aqueducts, the Coliseum and the Pantheon, is so impressive that most people simply reject the notion of Roman technological inferiority out of hand. And it is of course true that in a society with hot and cold running water, an agriculture sufficiently productive to feed the population and a wealth of military machines to make them invincible on the battlefield, there was no keen desire for earth-shaking new inventions.

A fine example of Roman engineering is Julius Caesar's bridge over the Rhine of 56 BC. This alone should suffice to make the point that it was not lack of technical sophistication that held back the Romans and their predecessors in the Mediterranean world from entering the industrial age.

Caesar's soldiers built this wooden bridge, 7–9 meters wide, in 10 days so that his entire army could cross the river to punish some Germanic tribes for crossing into Gaul. The breadth of the river at this point (probably situated near the modern city of Koblenz), 300 feet, and its great depth and powerful currents, made the task of building a bridge under field conditions that was big and strong enough for the entire army to cross a sheer impossible task. That this was nevertheless accomplished in 10 days illustrates the sheer perfection of Roman engineering even during a military campaign. According to his report of the Gallic campaigns, the future dictator himself designed the bridge and solved all technical problems, and this may well be true in view of his reputation as one of history's most talented individuals. In June 56 BC, he and his army marched to the other side. Having satisfied his honor by burning a couple of villages, he returned after only 18 days over the same bridge and had it destroyed behind him.

Probably the best illustration of Roman engineering and its impact on society is the system of great aqueducts bringing fresh water to the cities of the empire. After capturing the water of the springs high in the mountains, miles and miles of aqueduct channeled it down along underground passages, over ravines on top of tiered arcades, until it reached the massive reservoirs from which it served the public fountains, private residences and the many public bathhouses. The water flow was powered entirely by gravity with a gradient as small as 34 cm per km. To cross depressions, Roman engineers used inverted siphons to force water uphill driven only by hydrostatic pressure, without any need for pumping.

For Romans, baths were not a luxury but one of the foundations of civilization. Even the meanest citizen of Rome or other cities had free access to fresh, clean water and to a public bath. This is still evident from the ruins of the former public baths of the emperors Caracalla and Diocletian in Rome. In the former, a total of 1,600 people could take a bath simultaneously with as many as 3,200 in the latter. While Rome had more than 1 million inhabitants in its glory years, smaller cities, such as Pompeii with its 20,000 inhabitants, also had public baths. All these baths as well as many private residences had central heating. Also the sewer systems of Roman cities were generally excellent and it can safely be stated that the average New Yorker of today is hardly better off than his Roman or Hellenistic equivalent 2,000 years ago with respect to water, wastewater and sewer systems.

A very nice illustration of the aqueducts, how they functioned and what they meant for Roman citizens, can be found in Robert Harris' novel *Pompeii*[143]. One of the secrets behind these and other architectural marvels was cement. Although the Romans did not invent cement, they made abundant use of so-called pozzolana, a mixture of lime and volcanic dust found in southern Italy around Puteoli. This type of cement would even harden under water and it was the basis of the aqueducts and many a Roman monument, most notably Rome's Pantheon (around the year AD 120), which still has the largest dome ever supported by masonry.

In spite of its success in building a wealthy and highly civilized society, Rome's age of empire in the Mediterranean contributed little to technological progress and never reached the same level of technological sophistication as China. Apart from slavery, which may have been a major factor inhibiting innovation, certainly the geographical location was a factor. While the empire included Egypt and Mesopotamia, where irrigation continued to generate enormous surpluses (as it would still do in the Moslem Empire centuries later), in essence Roman society was very much a subsistence economy. This was not entirely due to technological deficiencies, but certainly also to the much lower productivity of the lands in the Mediterranean Basin (apart from the fertile river valleys allowing irrigation agriculture) as compared to China and Western Europe[144]. Without the surpluses from the alluvial river plains, with the continuing flow of grain from provinces, such as Egypt, Roman society would have looked very different.

We do not exactly know the life of the average citizen, but can get some glimpses of it from literary sources such as Pliny's letters. As mentioned, most farming was practiced on large estates owned by the magnates and often worked by slaves. They produced for the markets. We have already seen that in China most peasants owned their land, and slavery did not occur. But rich Chinese also let their land out to tenants and this also happened in the Roman Empire at the time of Trajan (AD 98–117) although not frequently. Pliny was one such landowner to let his land to tenants. In one of his letters, he describes the difficulties he has in collecting the rents and the lack of interest of his tenants to reduce their debt (they were often in rent arrears) because they had no hope of being able to pay off the whole.

143 Harris R. *Pompeii: A Novel*. Ballantine, New York, 2003.
144 Braudel F. *Memory and the Mediterranean*. Knopf, New York, 2001; Diamond J. *Guns, Germs and Steel*, Norton, NY (1997).

He then comes up with the idea to take a fixed share of their produce rather than a money rent[145].

Pliny's letter does not suggest that small-scale farming was a very lucrative business in the empire. Another example of that is found in *The Golden Ass*, a novel by Apuleius (125–180) which relates the adventures of Lucius, who experiments with magic and is accidentally turned into an ass. Falling into the hands of a poor market gardener in Greece who sells much of his produce every day on the local market, Lucius soon realizes that agriculture in the Empire is not a very profitable business. The gardener was clearly not very well-off, living in a shed and eating his own old lettuce in winter when there was nothing to sell[146]. It is likely that most small farmers used their harvest mainly to support themselves and had only meager surpluses to sell in the cities for cash to pay their taxes and buy the personal items they desired and could afford.

In spite of inefficiencies in such major areas as agriculture and metallurgy, the Empire was a brilliant society of cities with a thriving middle class and a monetary economy where everything could be had for cash. The extensive trade made possible by the Pax Romana greatly facilitated economic growth. The art world flourished and it would have occurred to nobody that Roman society was not the envy of the world. There would have been no inclination to change it significantly. Yet, eventually it did collapse. First, it lost its western part in AD 476, leaving the eastern half to survive almost 1000 years longer. Then, after a brief return to great power status in the 10th and 11th century, the Empire irreversible declined after in 1204 its capital Constantinople was captured and sacked by the Western crusaders, their co-religionists on their way to fight the Moslems. Managing to come back yet again, the last Roman emperor, Constantine XI, now only ruling over small parts of the Balkan Peninsula and Asia Minor, was killed defending his city against the Turkish onslaught of 1453. The most stable and successful empire ever, but almost without a single macro-invention!

As in the case of China, historians have great difficulty seeing the cause of the decline and fall of this great empire simply for what it was: successful encroachment by foreign people who were militarily more successful. In his masterpiece *The Decline and Fall of the Roman Empire*, the British historian

145 Radice M (translator). *The letters of the younger Pliny* (3, 19). Penguin, New York, 1963.

146 Graves R (translator). *The Golden Ass*, Noonday, New York, 1951.

Edward Gibbon (1737–1794) considered the decline a consequence of internal rot: a loss of manliness and the will to fight among the approximately 65 million inhabitants of the empire. Gibbon thought the rise of Christianity, which had become the state religion in the 4th century, a main culprit of Rome's fall[147]. Christianity, according to his reasoning, was sucking up resources without giving anything productive in return. This is based not only on the substantial gifts to the church but also on the many men and women withdrawing from society to become monks or nuns thereby no longer contributing skills or revenue to the empire. Other facts that have been considered are the frequent civil wars and the high tax pressure which led to a chronic lack of financial resources. However, is it true that the Empire fell because of a gradual process of internal rot?

As we can conclude from Figure 2, inventiveness in the lands of the Roman Empire dried up early in the first century, well before symptoms of decline began to appear, i.e., hundreds of years before the Empire lost its Western European territories in the 5th century AD. Therefore, a lack of technological progress was certainly not a cause of the fall of empire, although progress in military technology might have prevented it.

Like most other great states, Rome knew its ups and downs and especially in the 3rd century the Empire suffered from internal conflicts and frontier wars. But a series of strong emperors, demonstrating that even an age-old, non-innovative empire could still respond creatively when the need was there, restored peace, stability and prosperity. They re-organized the military, implemented a more centralized bureaucracy and re-structured the tax system resulting in the increased revenues necessary to protect the citizens of the empire. And it worked! There is now mounting evidence that instead of declining, prosperity, agricultural productivity and wealth generally increased in large parts of the empire, most notably in Asia Minor (modern Turkey) from the third to the seventh century. Also North Africa was doing well, at least until the invasion of the Vandals in 430. And if there could be any doubt about the continuing proficiency of the empire's craftsmen and architects, a visit to the never-surpassed monument of Roman ingenuity, the 6th century church of Hagia Sophia (Ayasofya in Turkish) in Constantinople (modern Istanbul) will wake you up.

147 Gibbon E. *The Decline and Fall of the Roman Empire*: Volumes 1-3. Knopf, New York, 1993.

Hence, it is simply not true that slow economic and social erosion was gradually driving the once proud Roman super state over the edge. On the contrary, were it not for an unfortunate combination of outside threats there is a fair chance that the *Pax Romana*, considered by the father of the 'internal rot' theory, Edward Gibbon, as the happiest period humanity had ever witnessed, would still exist today in the Middle East and parts of Western Europe.

According to Peter Heather in his book *The Fall of the Roman Empire*, there was no sign in the 4th century indicating the imminent collapse of the Empire's Western European territories[148]. It was the military superiority of barbarian tribes, most notably the Huns, which ended the Roman Empire in the West. This all happened in a very short time span, i.e., by around the year 400. The Huns were driving other peoples, including Germanic tribes, before them. In what was really an unfortunate series of events, the western part of the empire as well as Africa were invaded by various tribes who settled there permanently. The usual internal strife prevented the necessary vigilant response and although the Romans, as they had so often done in the past, could still have prevailed, this time it was not to be.

While the eastern part of the empire still made a partly successful attempt to recover the lost territory, Western Europe was permanently lost and somewhat later, in the 7th century, Africa and Syria were captured by the Arabs (see Chapter 10). Collapse had nothing to do with loss of inventiveness and was certainly not a long drawn-out process of rot. Indeed, a state can flourish without further additions to its technological repertoire! But loss of inventiveness is not without consequences.

First of all, in the absence of technological progress there are no longer increases in wealth. This is especially a problem when spending needs increase, as they did in the Empire due to increasing amounts of money going to the army and (later) the Christian Church. Without technological progress to keep pace with increased spending, more contributions were demanded from the weakest segments of society, such as the remaining free peasantry. The latter increasingly became bound to the land and a mere work tool for the great landowners and the state. As we will see in the next chapter, this situation was the beginning of the manorial system that became the heart and soul of the European Middle Ages.

148 Heather PJ. *The Fall of the Roman Empire*. Oxford, New York, 2005.

Initially, compensation was found in greatly expanded trade, which was made possible by an increasingly safe and efficient infrastructure, mostly involving improved shipping but also roads. Eventually, however, there were no longer increasing benefits from trade alone. Hence, it has often been argued that with the ever increasing military expenditures, lavish gifts to the church, the reluctance of the great nobles to pay taxes and a desperate peasant population who could no longer pay the increasing amounts of taxes, the system collapsed under the weight of its own financial needs.

The point has been made that from the 2nd century onwards Roman society became an oppressive police state. And it is true that state control increased tremendously. Similar to the situation in China under the Ming and Qing dynasties (and in our own 21st century), we see a growth in bureaucracy with increased regulation. This went so far that under Emperor Diocletian, occupations, trades, and professions became hereditary. He forbade a man or his children to change jobs!

But all these crushing taxes and draconian decrees notwithstanding, most citizens must have appreciated the increased comfort, peace and security they got in return. The same Diocletian, who ruled from 284 to 305, was a brilliant leader and his reforms restored the political and economic system. Although there is ample evidence that taxes increased over time, especially when in the 2nd century Rome had to face not only northern barbarians but a reinvigorated Persian state, there is little evidence for an economic collapse. Indeed, as in China, taxes were relatively mild as compared to the demands of citizens in our modern era. Problems only occurred in times of disaster when farmers were simply unable to come up with even the smallest of a contribution.

Overall, therefore, there is surprisingly little evidence for a collapse of the Roman Empire as a consequence of internal rot. In fact, the late empire did very well. But what we do see, however, is a loss of incentive to vigorously participate in the political arena. Like in Ming China, there were now many opportunities for indulgence in the pleasures of life, including intellectual growth. Literacy was more widespread than ever before. There were some great thinkers, such as Tertullian (160–220), Plotinus (205–270) and Augustine (354–430), but also a great many romantic novelists. The romantic novel in the late Roman Empire was as typical a manifestation of the culture of the time as the multimedia player in the 21st century. Works by novelists as the aforementioned Apuleius (125–180), Xenophon of Ephesus (2nd or 3rd century), Iamblichus Chalcidensis (245–325), Achilles Tatius (late 2nd century), Lucian Longus (3rd century) and the author of Daphnis and Chloe, Heliodorus of Emesa (3rd century), were widely read[149].

What we basically see, therefore, is a highly successful late-Roman society with an advanced governing bureaucracy that managed to keep the peace and provide ample opportunities for their subjects to live a decent life without too much interruption by violence or environmental disasters. Similar to the Qing state in the early 19th century, the late Roman Empire did not collapse from internal problems. It was due to external blows in combination with a chronic shortage of men willing to lead that ended one of the most successful societies the world has ever seen. Like the Qing, the Roman Empire would still exist were it not for the unfortunate series of events described above over which they had limited control.

In the opening scene of the movie *Gladiator*, directed by Ridley Scott, we see how the impeccably organized Roman legions defeat an army of disorganized rabble, ending a long war on the Roman frontier. The scene depicts the end of the longest war of the Roman imperial period, against the Germanic Marcomanni. In contrast to what some may think after having seen the movie, this was not a typical war of a vastly superior power against a backward society, such as the wars fought by Britain against the Zulu's in Africa far later. Although as warm-up acts we witness an impressive Roman bombardment of their enemies by an arsenal of catapults and the famous tortoise formation of Roman infantry slowly marching forward, the actual battle is still a very primitive man-to-man fighting with little evidence for Roman technological superiority. The great difficulties of Emperor Marcus Aurelius subduing these barbarians are clear to everyone visiting the Roman forum admiring the Column of Marcus Aurelius, depicting these Marcomannic wars. He would never have expressed such pride in the accomplishment when this was just a simple expedition against a primitive rebel enemy.

The reality was that Rome (like China) simply lacked the military technology to annihilate the barbarian tribes that constantly threatened them. Although better organized and usually wearing protective armor, the Roman military were not superior to their Germanic opponents. And by that time the Germanic people were no longer as primitive as when Julius Caesar first encountered them, 200 years earlier. Their metallurgy was no

149 Grant M. *The Climax of Rome*. Weidenfeld, London, 1968.

longer inferior, their agricultural practices so greatly improved that their numbers began to increase substantially and their societies more structured and organized than before. Fortunately, Roman diplomacy could usually avoid costly wars against these people. But sometimes, like the Chinese, the Romans had to fight a superior military force in the form of steppe people.

Unlike the Europeans of the future, both Romans and Chinese lacked the machine guns that would have given them an edge over the vastly superior cavalry force some steppe people could bring to bear. One of these steppe people, the Huns, migrated from the Eurasian steppe to the Roman eastern border lands. Steppe warfare as practiced by the Huns and later the Mongols was simply irresistible. The horse was the tank of the ancients and extremely effective, especially with the use of stirrups and special saddles allowing their riders to use their bow and arrows to hit targets with a superhuman precision from a safe distance while running at a full gallop. The Chinese never had an answer to that, probably also because the military was not the most fostered part of Chinese society. This is the simple explanation for how such highly developed states with almost 100 million citizens could ever be overrun by nomadic tribes.

In the Roman Empire the military were the most powerful segment of society. They learned from the Huns' military predation upon their state and developed a mobile cavalry force themselves. Unfortunately, by overturning the balance of power in the Roman northeastern border lands the Huns were driving Germanic Goths into the empire. This led to the dramatic Roman defeat at Adrianople in 376. Still, a combination of Roman military skill and diplomacy could have saved the day, especially since the Huns were not interested in permanent occupation and the Goths as other Germanic people were solely interested in permission to settle on Roman lands and participate in their successful society. It really was an unfortunate combination of internal strife and external pressure that eventually led to the collapse of the Western part of the empire in the 5th century. This came as a surprise to everyone and it is certainly not true that internal rot was so obvious that everybody was waiting for the end as is sometimes suggested by those who point to internal decline. The Roman Empire in the West collapsed because it simply lost one battle too many.

The Eastern part was more fortunate. It was more populous and much richer than the west, allowing it to better maintain its defenses. For example, Egypt remained an indispensable source of corn for the Eastern Roman

Empire until it was captured by the Arabs. The Eastern capital, Constantinople was the largest and richest urban center in the Mediterranean during the late Roman Empire, mostly as a result of its strategic position commanding the trade routes between the Aegean and the Black Sea. Also militarily it was strategically well located and virtually impregnable, defying assaults for almost 1,000 years. This may partly explain why the Eastern Empire managed to hold out so much longer. In the west the Roman Empire fell in the 5th century and made place for a loose constellation of semi-barbarous states that would eventually become Europe.

Chapter 10. Against All Odds: The Rise of Europe

The French historian Fernand Braudel (1902–1985) called Europe an "Asian peninsula, a little cape linked with the East by a broadening continental land mass"[150]. Europe had shared in the fruits of the Neolithic revolution and was populated by peaceable villages and hunting tribes until overran by Celtic invaders in the 4th and 3rd centuries BC, as a late part of the Indo-European invasions. After Julius Caesar in the first century BC conquered Gaul, which basically included modern France, Switzerland and Belgium, later also parts of Britain, Holland, South Germany, Austria and Hungary were included in the Roman Empire. In the 5th century parts of Roman Europe were overran by barbarian Germanic tribes, such as the Saxons in Britain, the Franks in Belgium and the Goths in Spain and the south of France. Barbarian Europe was a fact after Odoacer, a Germanic chieftain, formally ended the reign of the last emperor of the West in 476 to establish a kingdom in Rome's original heartland. Until that time, Europe had contributed nothing of significance to technology and civilization, and until the end of the 10th century it was one of the most backward parts of Eurasia. Yet, the seeds of its future greatness were already there.

First of all, the European climate was optimal for agriculture. With its soft winters and not too hot summers, water in all seasons, extensive hardwood forests and fertile clay soils, it was waiting for the necessary tools to optimally exploit all this silent wealth. Tools to break open the sturdy

150 Braudel F. *A History of Civilizations*. Penguin, New York, 1995.

ground or cut the heavy trees were simply not there because the Romans were used to agriculture in the Mediterranean Basin, which was easier but far less productive. Even agriculture in ecologically diverse China with its river valleys was not as robust as that in Europe, albeit much less fragile as in the Mediterranean[151].

The factor that did most to help Europe climb the ladder of technological progress is geography, which did not lend itself well to forced unification. Indeed, the Mediterranean united the Roman Empire, and China's heartland is linked together by two navigable river systems. However, while Europe has plenty of rivers, none of them interconnect more than a fraction of its total landmass. The same is true for its coastline. While important trade links, the Baltic, the North Sea and the Atlantic Ocean are far from the equivalent of the Mediterranean. Europe's coastline is also not smooth and differs from China also in that respect. Its many peninsulas almost guarantee political diversification and England's insular status was considered the key to its 'splendid isolation'. Hence, the geographical situation does much to explain why, in the absence of a foreign unifier, Europe's multistate system would always remain. And this would prove to be critical in the fierce inter-state competition that like China's Warring States and the pre-Roman Hellenistic states also in this case would lead to technological advances and therefore great wealth. But this would be the future. The reality of AD 476 was much less auspicious.

That Europe survived more or less as an independent entity is a miracle by itself. Even before the formal end of the Western Empire an improvised allied force under command of the Roman general Aetius barely managed to defeat the Huns, one of the many nomadic tribes of the Asian steppes, at Châlons-sur-Marne. Soon Europe was attacked from all sides. In the South, Arabic troops in the wake of the Prophet Mohammed's revelations had overrun the possessions of the Eastern Roman empire in Africa and the Middle East and besieged the capital of Constantinople in 719. Proving impregnable, the city that controlled the important connection between the Mediterranean and the Black Sea kept the Roman Emperor in charge of his now much diminished empire and saved Western Europe from an attack through its soft underbelly. Nevertheless, the Arabs founded a sparkling new empire on what once were the Syrian and African possessions of the Roman Emperor. The Moslem armies even destroyed the West-Gothic

151 Diamond J. *Guns, Germs, and Steel.* Norton, New York, 1997.

kingdom in Spain and invaded France. Charles Martel, the leader of the dominant barbarian kingdom in the West and Charlemagne's grandfather, withstood them with difficulty at Tours in 732.

For a short while it seems that under the strong leadership of Charles Martel and his successors, Europe was climbing out of the dark ages uniting Europe under the new Roman Emperor, the Frankish King Charlemagne, crowned in 800 by Pope Leo III. This period, from the late 8th into the early 9th century is known as the Carolingian renaissance.

But soon new hordes of invaders presented themselves. They came from the Scandinavian and Slavic lands in the North and North East, from the South where Moslem raiders had occupied the main islands in the Mediterranean (including Crete and Sicily) and even held some passes over the Alps and coastal areas in the South of France, and from the East in the form of the Avars and Magyars, tribes related to the Huns. The Mediterranean was now a Moslem lake, effectively cutting Europe off from any form of trade with its more wealthy neighbors. Indeed, for hundreds of years commerce ceased to be one of the branches of social activity, apart from some very limited trade mainly by Jewish merchants who reached western and northern Europe through Spain.

In the initial turmoil of continuing barbarian invasions and increasing violence during the end-days of the Roman Empire in the West, agricultural productivity declined even further and the sparse records show that technology-wise there was indeed very little to choose from. Most tools were now made of wood, and iron became even rarer with few people to work it. Population density had also reached lows that had not been seen since the first Neolithic farmers began to populate Europe, which was now a profoundly uncivilized place with the average peasant certainly worse off than most of the ancient hunter-gatherers of the Stone Age. Even their relatively rich lords would have made themselves the laughing stock of many a noble court in the Middle East or Asia. Early medieval times are called the 'dark ages' for a reason!

Thrown back on itself, without access to the Mediterranean, how then did Europe eventually manage not only to get back on its feet but develop into a technology powerhouse and the strongest and most sophisticated economic system the world had ever seen? Well before the year 1000, which is generally considered as the start of one of the most dramatic technology explosions in history, the tide had begun to turn. Many Scandinavian raiders established themselves peacefully in various parts of Western and Southern Europe, becoming a part of European society. The remaining Vikings in Scandinavia became traders heralding the Baltic Sea trade and foreshadowing the Hanseatic League, an economic alliance of trading cities along the coast of Northern Europe. A symbolic watershed was the battle of Lechfeld, near Augsburg, when in 955 one of Charlemagne's successors, Otto I, in the lands which now comprise Germany gained the upper hand over a fresh host of Magyars that had penetrated Europe. This established Otto as the leading monarch of the day[152].

The key factor in the turnaround was the significant change in the power structure in what used to be the western part of the Roman Empire. The Germanic invasions of the 5th century now forced the remaining free farmers to place themselves under the protection of one of the magnates. They became 'serfs', holding land from the lord of an estate in return for fixed dues in kind, money and services. This system of independent economic units with essentially only two classes, i.e., lords and serfs, is called the manorial system. It was not new and had gradually come into place in the Roman Empire at least from the 3rd century onwards, as we have seen. But with the empire still in existence, central control with its bureaucracy and tax collection offered little incentive to innovate. This changed after the Empire had collapsed. First, the change was for the worse. The peasants under the manorial system no longer produced for the market, which no longer existed with the virtual disappearance of the cities, and most of the technology that was there disappeared or greatly diminished[153].

But something else had now also changed. In striking contrast to the barbarian invasions in China, which always resulted in the adoption by the conquerors of Chinese civilization and administration and a seamless takeover of central government as a new dynasty, nothing of the kind happened in the lands of the Western Roman Empire. Part of the explanation for this may be the much lower level of sophistication of the Roman civilization than the Chinese, but the key factor is most likely the lack of individual strongmen among the barbarian tribes. They simply lacked a brilliant leader like Genghis Khan (whose power went undisputed during his reign) and

152 Whitton D. "The society of Northern Europe in the high Middle Ages". In: *The Oxford History of Medieval Europe* (editor: Holmes D). Oxford, 1988.

153 Heer F. *The medieval world: Europe 1100-1350*. Weidenfeld, London, 1961.

his grandson Kublai Khan, who finally took over all of China and founded the Yuan Dynasty.

No new dynasties for the Germanic conquerors, most of whom peacefully coalesced with the Roman magnates to participate in the manorial system which soon became the dominant system everywhere in Europe, from Britain and the Low Countries to Eastern Europe and Italy. However, the upper layer of society was no longer the same educated group of Roman magnates firmly controlled by the emperor. While they could not have made up more than 10% of the original Gallo–Roman population at the time, the Germanic conquerors managed to profoundly alter the war-averse state of mind that had become so characteristic for the citizens of the Roman Empire in the glory days of a well-structured, successful society. From the sources, most importantly Gregory of Tours and his *History of the Franks*, it is clear that early European society was very violent[154]. War was endemic and not only because of continued invasions of new barbarians. Internal strife was also plentiful, and it is not hard to see how the cream of this society developed itself into what we now know as European knighthood. This European warrior cast was a product of the feudal system as this developed itself in the absence of a central power system. Europe fragmented into many semi-independent mini-states, run by noblemen, including descendants of former Roman magnates. They held their lands from a lord to whom they had to pay homage and provide military service. This could be the king or another nobleman who in turn then had to pay homage to the king. These kings were not a shadow of the former Roman Emperor with his professional servants and standing army. They also had no means of taxation.

Naturally, knights were not particularly interested in agricultural management. As long as the serfs provided them with the required part of the harvest, they would leave them to their own devices. Feudalism in early Medieval Europe was therefore much more productive than the previous practices of either slavery or tenants controlled by the magnates (who had little else to do than supervising them). In the absence of their masters, who were almost non-stop involved in some of the many chronic conflicts of the time, peasants were now seriously interested in increasing production and that would soon show. Of course, in China feudalism was abandoned early on and replaced by a system based on free peasants. This would eventually happen in Europe also. The difference underscores how much further ad-

154 *Gregory of Tours — The History of the Franks*, Penguin, New York, 1974.

vanced China was as compared to Europe, which would catch up no sooner than the 18th century.

While as mentioned above the European climate was optimal for agriculture in terms of temperature and water, its extensive forests and heavy clay soils resisted the limited agricultural tool box of the times. The Romans did not have a lot of iron tools and their wooden scratch plow was entirely unsuitable to work the hard soils in most of the Northern areas. These technological limitations explain why the Romans never penetrated in the heavy forests of Germany or the clay soils of the North. Instead, their farms were restricted to the hilly areas and loose soils in Gaul or Britain. While no match for agriculture in many parts of the eastern empire, the meager surpluses in the west nevertheless added up and together with the Mediterranean trade provided sufficient wealth for the magnates not to seek too hard for productivity increases. Although Europe had the geographical conditions to become an agricultural success story, technological backwardness due to a lack of incentives had always precluded it. From around 700 onwards, however, things began to change and the next few hundred years would witness an increase in agricultural productivity the Romans could never have imagined.

To gain access to northwestern Europe's fertile but heavy clay soils the peasants began to make more frequent use of the heavy wheeled plow drawn by oxen, which was capable of breaking and turning the sod. Since individual farmers could not afford a plow of their own, agriculture was a communal activity. A second improvement was the adoption of a three-course rotation system to replace the classical two-course rotation. This consisted of sowing a spring crop (oats, barley, peas or beans) harvested in summer, wheat or rye in autumn (harvested the following summer) and a year of fallow to restore fertility. This system was in general use in Europe by the 11th century. (Crop rotation and multiple cropping were used much earlier in China.)

A third improvement was the use of horses as draft animals, both stronger and faster than oxen. This could not be done before the introduction of the horse collar and horse shoes in the 10th century. (As we have seen, a horse collar was already in general use in China at least 500 years earlier.) From now on innovations in Europe would continue to be made, many of which were a result of improvements in metallurgy. Iron smelting (except in China) was limited to bloomeries and required repeated hammering to create tools. Soon, every village had its blacksmith. Bloomeries became progressively larger with the first blast furnaces emerging in the 15th century. Iron began to be used much more frequently, for example, in critical components of the plough, i.e., coulter, share and mould-board. Mining, metal working and manufacturing, especially of woolen cloth, were taking great strides. Machines became a common sight in the landscape of medieval Europe. Water mills, wind mills and tidal mills converted the power of the elements into work[155].

Water mills were used by the Romans as well as the Chinese at least since the 1st century BC. Initially not more powerful than a donkey mill, they were improved to about 3 horsepower. These machines were described by Vitruvius the author of the famous treatise *De architectura*[156], written around 27 BC. Book 10 of this treatise dealt with mechanical engineering. However, the Romans were no match for the Europeans, who built many more water mills than the Romans ever did, greatly aided by the fortuitous wealth of rivers and streams in Europe flowing throughout the year. Although we do not know much about the early centuries after the fall of the Western Roman Empire, we are quite well informed about the situation since the turn of the millennium. The best example is the English Domesday book, a survey commissioned in 1085 by William the Conqueror (who conquered England in 1066) containing records for 13,418 settlements in the English counties south of Scotland[157]. For a population of roughly 1.4 million at that time, there were 5,624 water mills. Over the next centuries this number would grow dramatically.

Water mills were built along the banks of a river or midstream. Sometimes rivers were dammed to provide a fall of the water that was sufficient to drive the mill. The mills were used to grind the grain or crush olives (in the south, to make olive oil), but also for other purposes. For example, by using cams projecting from the axle of a waterwheel they could turn the rotary movement into a reciprocal motion. This gave rise to the trip hammer. Such trip hammers were already operated by the Chinese around AD 300. Water-driven hammers were used for tanning (processing leather), fulling (shrinking and thickening of cloth), making paper (by pulping rags), and forging iron. Iron ore was more common in Northern Europe than in

155 Gimpel J. *The medieval machine: the industrial revolution of the Middle Ages*. Holt, Rinehart and Winston, 1976.

156 http://www.vitruvius.be/

157 http://www.domesdaybook.co.uk/

the Mediterranean and iron began to be used in the European Middle Ages on a much greater scale than ever before, except in China. Iron smelting was improved by using charcoal-fired furnaces with forced draught provided by bellows, also driven by water power. Eventually this led to the blast furnace, almost two thousand years after its first appearance in China. But in Europe the pig iron tapped from such blast furnaces was turned into the more malleable wrought iron by repeatedly heating and hammering to remove excess carbon and silicon. This was done by the trip hammer in a finery forge.

In their eagerness to harness as much energy as they could, the Europeans also invented tidal mills, not used in antiquity. They were built in low lying areas where the river flow was weak, but would never become as important as water mills. Wind power, by contrast was important. Both horizontal and vertical-axle wind mills were invented in Persia in the 7th century and from there found their way to Europe and China during Mongol times. European, vertical-axle wind mills were first constructed in the mid-twelfth century (about the same time as in China). They were especially popular in the north since they could also operate under freezing conditions. In the 13th century they were introduced in Holland — one of the many feudal mini-states that eventually fell under the hegemony of the Spanish Habsburgs — to drive the pumps that kept the water out of their diked lands known as polders. They began to use them on a large scale to reclaim land from the sea, a sure sign of the beginning of overpopulation.

Clearly, the peasants under the manorial system managed to show significant individual initiative. While the lack of documents from those early times essentially constrains a complete understanding of this phenomenon, we can at least speculate. First of all, after the 8th century we see an increasing level of stability centered round feudalism and the manorial system in which every person had a place. It was the Church who enforced its spiritual will on the entire population of Europe, convincing the nobility that peasants had rights also. The Church also played a more direct role in Europe's economic miracle. St. Benedict of Nursia, who died in 543, introduced a 'Rule for Monks', not based on extreme asceticism but on physical labor, prayer and healthy living. This rule proved to be eminently successful and by the end of the 9th century the Benedictine rule was the basic rule for all monasteries in Europe except Ireland.

The monks would not only preserve Roman culture but also become a major driver of what would become the first medieval agricultural revolution. Monasteries as well as the secular clergy had large estates often obtained as gifts from pious lords or kings. The Benedictines ran their estates more efficiently than the lay nobility, pioneered agrarian science, and helped to lay the foundation for Europe's increasing reliance on machinery. Importantly, many peasants were essentially free to run their operations as they saw fit. As long as they provided their lord with fees and taxes (in the form of foodstuffs) and labor service, he did not care exactly how they ploughed their land.

Under such a system of mutual rights and liberties, the peasants now certainly had incentives for improvements. Apart from the spiritual duty to work their land the best they could, aggressively supported by a church which was now highly active in improving agriculture (as they had never been in the Empire), there was their natural inclination to save their own labor and reduce its burden. They were phenomenally successful, as can be judged from a dramatic increase in the European population. Shortly after the fall of the empire in the west there may have been as little as 10 million people in Europe. Apart from the endemic violence the disaster was made complete by an outbreak of bubonic plague in the 6th century that was the worst until the one that began in 1353. From this low the population would begin to rise in the 7th century to an estimated 20 million people in the year 1,000 and 50 million around 1300. This is still only half the population of China at the same time, but the growth rate is highly impressive. Indeed, the population of Gaul and Britain hardly increased under the Romans and actually decreased quite substantially from 200 A.D. onwards.

The excess population resulting from agricultural improvements since the 7th century had to go somewhere and many of them went to find new land. Around the turn of the millennium we see an enormous increase in new land that was brought under cultivation. Those were either free peasants, trying their luck elsewhere, sons of serfs who did not inherit any land, or serfs that simply ran away. While the lord could try to get them back, once far away from the manor they became anonymous and in a time when manpower was short nobody asked questions. Since there were plenty of landowners willing to provide strips of land to tenants to cultivate the former wilderness areas of which there were still plenty in the sparsely populated Europe of the early middle ages, the number of free peasants was increasing steadily.

Improvements in agriculture slowly led to wealth increases and the renewed usage of money in the estates. Although money had never completely disappeared, the manor had no need for it because it was completely self-supporting. This was necessarily so because trade and with them cities and artisans had virtually disappeared. With the increase in agricultural wealth and the gradual stabilization of European society commerce slowly came back. It had never disappeared entirely. Some maritime cities of Italy, most notably Venice, continued their trade with Africa and Syria, selling young Slavs (hence, our word 'slave') to the Arab world as well as timber and iron. Pavia was the early hub in Northern Italy where the trade routes from Northern Europe found their Venetian partners for exchanging goods.

As we have seen, the Vikings substituted trade for piracy in the 10th century, which opened up the coast of the North Sea and the Baltic. Merchant associations within the cities of Northern Germany, the Low Countries and the Baltic formed the Hanseatic League, which maintained a monopoly on trade over the Baltic and the North Sea from the 12th century. Their cargos consisted of salt, herring, grain, timber, honey, amber, ships stores, and other bulk commodities. Europeans now produced a new and innovative ship design, the so-called cog ship. Cogs were larger, wider and higher than the long, narrow Viking vessels or the Mediterranean galleys, cheaper to build, more seaworthy and could carry a lot more cargo. These ships, high in the water and with rounded sides, were used everywhere in Europe by 1300.

A key export article for Europe was cloth, with its undisputed center in the Low Countries. The land there at the mouths of the Rhine and Meuse was highly suitable for sheep farming and once this supply of wool was exhausted, Flemish agents began to import it from England. This is the time when cities such as Ghent and Bruges developed into the centers of wealth still recognizable today. Originally, cloth was exported by the land route over the Alps to the market in Pavia. However, once the Normans together with some enterprising Italian cities, most notably Pisa, had driven the Arabs from the Tyrrhenian Sea, trade intensified dramatically. Around 1300, Venice and Genoa began to organize fleets to Flanders and England.

Former peasants or landless sons of peasants now came to the cities, which were mainly the seats of the bishops, to live under their walls and become so-called burghers to work for merchants or set themselves up as artisans. In turn, these reinvigorated cities again became a market for their fellow agriculturists. Food as well as tools could now be bought again on the market, which heralded the end of the manorial system and its serfdom. Most lords now decided to forego the labor services and instead asked for a cash payment from their serfs and tenants.

After a very slow and gradual beginning we therefore see a whole new society arising, a society that in many respects resembled that of China 1000 years earlier. After its early feudal period, peasants in China were free during the Warring States period because it was found out that this gave rise to productivity increases, enhancing the power of the state. They practiced intensive cultivation selling their surplus on a free market. There is no doubt that also China's peasants had ample reason to innovate, although their circumstances were different from their European counterparts. In China it was the state which often protected them against unreasonable demands from the magnates (which remained a factor also in China). In Europe it was the Church which played that role, but later also the kings of the different nation states that began to emerge between 1000 and 1500.

Both in Europe and in China we see repeated peasant uprisings when liberties were taken away or tax burdens became unreasonably high, a sure sign that they had something to lose. By contrast, in the earlier Mediterranean empires there is very little evidence for peasant uprisings or for peasant rights for that matter. These empires relied on extensive state intervention to deliver the enormous surpluses resulting from irrigation agriculture. The Mediterranean lands were not very suitable for small-scale farming, which goes a long way explaining why we see no significant improvement in food production after about 3,000 BC.

It is interesting to note that in both China and Europe the peasantry was a respected class in society with liberties and rights. While no doubt exploited to the bone by the rich and powerful, they must have had sufficiently high levels of self-esteem and freedom to innovate, either on their own or in combination with the lords and managers of their lands. They could also run away, which they could not do in the Roman Empire. Crossing borders into other lands where the rulers had different interests and conditions were better was always an option. Especially in the northeast where German rulers made inroads into Slavic lands, new farmers were more than welcome. We have already seen how important a system of vigorously competing states is for inventiveness. Hence, the fact that Europe had all the natural conditions to prevent unification is almost certainly a major factor in its success.

Although technology development would be interrupted by periods of stagnation or even regression, most notably the outbreak of the plague in the 14th century, in which almost a third of the European population was killed, progress would not stop. Instead, it would continue to drive the economy and bring wealth to more and more people, eventually resulting in what we now call the industrial revolution and our modern post-industrial society. Before discussing this final episode in the world's technological development, we first need to discuss another major advantage Europe had over the great civilizations that went ahead of it. As a late-comer, it profited optimally from inventions made elsewhere, most notably Islamic civilization.

Moslem assistance

With so many in the West these days feeling that Islam is intrinsically backward, misogynistic and violent, it may come as a surprise to find out that the key aspects of the European heritage, such as capitalism, science and the Greek philosophical tradition were acquired from the Moslem civilization. From the 7th century onwards, Moslem domination of the Mediterranean led to the establishment of two main centers of Islamic culture that were easily accessible to Europeans: Spain and Sicily. While initially there were not many contacts, this would change around the turn of the millennium. This was the time when Europe was slowly taking back the Mediterranean, began to make inroads into Spain and conquer Sicily. It was from there and around that time that Europe began to borrow so many useful tools, concepts and ideas from Islam that one sometimes wonders if not everything that Europe had was in one way or the other obtained through the Moslem civilization[158]. The various crusades against the infidels in Spain and the Middle East merely increased this one-directional flow of technology transfer.

Much of what Europe acquired in this way was not originally developed by Arab society but came from elsewhere. The magnetic compass and paper, for example, came originally from China. The astrolabe, an instrument that enabled astronomers to calculate the positions of the Sun and prominent

158 Pickstone J. "Islamic inventiveness". *Science* 313, 47, 2006.

stars with respect to both the horizon and the meridian, was introduced to the Islamic world through translations of Greek texts and fully developed by Moslem scholars during the 9th century. The astrolabe remained essential for indicating time and latitude to seamen until the middle of the 18th century. Especially India was a rich source for ideas, products and technology that through the Arabs eventually ended up in Europe. Already in the 7th century, there were Moslem trading colonies on the Malabar Coast, and the next century saw an Arab invasion from Sind at the mouth of the Indus, spreading slowly towards the Ganges. The sultanate of Delhi was founded in 1206, eventually followed (in 1526) by the famous empire of the Great Mogul.

However, while it is certain that India greatly contributed to science and technology, Arabs themselves also contributed. In astronomy and mathematics, for example, Moslem scholars were especially creative. Much of their work centered round techniques intended to formalize the relationship between the motions observed in heaven and the motions of the components of the astrolabe. In particular this work led to the production of tables of trigonometric functions. The Arabs were also great clock makers (both water clocks and weight-driven mechanical clocks) and it is likely that Europe's eventual success in making advanced time pieces was originally driven by Islamic sources. Also knowledge of the properties of lenses, critical for making spectacles, came into Europe through Islam. The properties both of lenses and plane, spherical, and parabolic mirrors were known to the Moslem mathematician and natural philosopher Alhazen (Ibn al-Haitham) (ca. 965–1039). In Part V of his *Opus Majus*, Roger Bacon (1214–1292) discusses reflection and refraction and the properties of mirrors and lenses, making use of Alhazen's work and also that of Alkindi[159]. It is likely, however, that the actual application of lenses for vision correction was first made in Italy in the late 13th century, when we see paintings appear with people wearing or holding spectacles.

The most important gifts to Europe from the Moslem civilization were Hindu-Arabic numerals, algebra, and double-entry bookkeeping. As first introduced by the Islamic mathematician Al Khwarizmi, Arabic numerals provided a major advance over the cumbersome Roman numerals. The development of a convenient number system assisted progress in science, ac-

159 Bridges JH. *The 'Opus majus' of Roger Bacon*, Volume I. Adamant, 2005.

counting and bookkeeping. Key to this was the use of the number zero, a concept unknown to the Romans.

In a sense, Moslem civilization was a successor civilization to the ancient empires of the Mediterranean, starting with the Sumerian empire at the dawn of history. It was also an intermediary civilization between China, India, the Greek–Roman heritage and Europe. It controlled such major trade routes as the famous silk road to China. Its sea routes to the Indian Ocean were made famous by the "The Seven Voyages of Sindbad the Sailor", a tale that has become part of *The Arabian Nights*, a collection of Persian, Arabian and Indian folk tales handed down through several centuries[160].

The Moslem empire inherited the two great irrigation agricultural centers of the Tigris/Euphrates and the Nile and brought hydraulic engineering to new heights elsewhere as well, most notably in Spain. They also introduced new crops, some obtained from China or India, such as rice, sugar cane, cotton, citrus fruit and watermelon. As could be expected, this led to increased food production and an increased population with cities as Baghdad and Cordoba as its splendid centers. Still, it could not change the inherent deficiencies of Mediterranean agriculture with its relative lack of fertile soil and tendencies to salination and erosion[161]. Similar to their predecessors, Moslem rulers were able to handle this as long as they were firmly in charge. Unfortunately, similar to the situation in the former Roman Empire, there was internal strife and they were vulnerable to military attack from outside.

In the 13th century, Islam lost its leadership position. The open access to its territories from all directions made them vulnerable to barbarian invasions, just like the empires before them. We have already seen that thanks to its geographical position, Europe escaped this fate, probably also because they had never been rich enough to be of interest to other barbarians. The Moslem empire, however, was invaded from Africa by Berber tribes, by the Europeans in their crusades, by the Mongols and by Turkish tribes. All of them, except the Europeans, would stay and adopt the Moslem religion. The last of them, the Ottoman Turks, established themselves as the new great power in the Mediterranean with Istanbul, the former city of Constantinople, which they conquered in 1453, as their capital.

160 Haddawy H (translator). *Sindbad: And Other Stories from the Arabian Nights*. Norton, New York, 1995.

161 Jacobsen T, Adams RM. "Salt and Silt in Ancient Mesopotamian Agriculture". *Science*, 128, 1251-1258, 1958.

The new leaders of the Islamic world were not as enlightened as their predecessors in Baghdad. They tended to curb independent thinking and governed through an authoritarian bureaucracy. While brilliant from the outside and in fact highly successful in keeping a huge empire together until 1918, this was a colossus on feet of clay. They were neither particularly rapacious nor violent — under the Turks there were no pogroms and they behaved extremely benignly towards their subjects of other religions — but their system lacked the opportunities of their neighbors in the West. Under the circumstances there was simply not a lot they could do against the ongoing environmental destruction undermining the once so spectacularly productive irrigation agriculture. Trade, always a major driver of economic growth, was constrained by strong competitors in the West, essentially closing off the Western part of the Mediterranean and competing with them in India and China after Vasco da Gama had opened up the Atlantic sea route round Africa (Chapter 11). Turkey now became one of those parts of the world unfavorably compared to a rising Europe, well on its way to world dominance. The reasons for Europe's success were the same that had previously led to the technologically competent empires of the Mediterranean, India and China, i.e., optimal geographic conditions and a system of strong states vigorously competing for dominance. This time it would be the most spectacular outburst of technology the world had ever seen.

Chapter 11. Europe and the Industrial Revolution

Italians are generally not considered a martial people. Yet its soccer matches sometimes allow us to peep into earlier times when violence and cruelty were the norm. For example, go and see a match in the Serie A between A.C. Siena and Fiorentina of Florence. Don't be surprised if, when Siena scores a goal, its supporters chant 'Montaperti!' to remind the opponent of their victory almost 800 years ago in the bloodiest battle of the Italian Middle Ages. In the Renaissance the rivalry between these two neighboring cities was notorious and they fought many battles. But the many other city states in the Italy of the 13–15th century were also at each other's throats almost all the time. This is illustrated by the saying, "better a death in the house than a Pisan at the door," which is attributed to whichever city was at war with the city of the leaning tower.

The incessant wars resulting from this fierce competition infested the country, leaving devastation in its trail. Fighting was literally non-stop with most cities employing professional armies to serve their cause. Nowadays we would immediately associate such bellicosity with disaster and economic devastation, but certainly not with the wealth and glamour of the Italian renaissance. But, in fact, the spectacular political, economic and cultural accomplishments that would fuel European civilization for centuries was born from the tensions inherent in the political system of free, intensely competing city states. It was rivalry and political chaos, not stability and good governance on which Europe's march to worldwide dominance was founded.

It was in the tumultuous times of the 13th to the 15th century that Italy astonished the world with its brilliance and inventiveness. It was then, not now, that unparalleled works of art were created, from Dante's "Divine Comedy," the epic poem still widely considered the preeminent work of Italian literature, Brunelleschi's great masonry dome covering the cathedral of Florence and Giotto's masterpieces of painting, to Michelangelo's Sistine Chapel and Leonardo da Vinci's precocious inventions jotted down in his notebooks.

Less well-known, but at the basis of all the wealth enabling Italy's artistic output in those days, was the great technological progress. Italian innovations stretched from new business methods (Arabic numerals, double-entry bookkeeping) to the mechanical clock and the industrial production of glass and paper. While much of this was based on earlier Chinese, Indian and Arabic inventions, the Italians implemented greatly improved versions of these novelties on a large scale, thereby truly transforming society as great inventions always do. Most importantly, the Italians of the renaissance made the first meaningful advances in ocean navigation. Ultimately this would lead to the long ocean passages around Africa, to the new world and across the pacific.

Now known as the renaissance, the period from the 14th to the 17th century witnessed Europe's rise to becoming a major center of political experiment, economic expansion and intellectual discovery. While other civilizations had reached this stage well before them, Europeans would bring it to new heights. Politically, we see the rudimentary beginning of the nation state. Economically, we see an ever greater role of the market economy and an expansion of the trade routes. Intellectually, we see the beginning of science as it is still now practiced all over the world. During the next centuries Europe would become the single most important center of technological progress, a position it would only lose in our own days. This rise from obscurity to great wealth could not have taken place without the great increase in food production, which as everywhere else led to surpluses and the opportunity for enterprising individuals to busy themselves with other things, including trade, mining, manufacturing and science. This was a consequence of Europe's robust ecological environment, which provided them with fertile soil with ample capacity for regeneration, and the new opportunities for its peasants to innovate. But it was its access to the Atlantic that would ultimately provide them with more and better trade routes than any other country before them. Europe was also the first ascendant state system to freely profit from inventions made elsewhere, the advantage of being a late-comer.

After the great plague, which wiped out almost a third of the European population, technology development had slowed considerably but was otherwise not significantly impaired and the nascent European state system survived. However, when after a century of decline around the middle of the 15th century Europe's population was again growing, there was a pronounced shift underway from the Mediterranean towards the Atlantic seaboard in the principal centers of economic activity. The consequences of this were global.

The rapid development of technological changes in navigation and shipbuilding in the previous centuries led to the establishment of the European colonial empires, first by Portugal and Spain, then by the Dutch republic, France and England. Portugal pioneered the era of the great discoveries that would be such a major determinant of Europe's economic success. Europe adopted the magnetic compass from China, probably through the Arabs, together with other navigational instruments. They had also begun to design ships other than the oared galleys common in the Mediterranean, such as the Cog and its successor, the Holk. In the late 1400s, the Portuguese developed the Caravel to help them explore the African coast. This broad-beamed ship had no oars but could sail very fast with its two or three masts with square sails and one additional triangular sail. It could also sail well into the wind and was equipped with the sternpost rudder imported from China. Caravels were the first ocean-going ships in Europe; they were large enough to be stable in heavy seas and carry provisions for long voyages.

The caravel and its successors were key to Europe's coming dominance on the high seas, in combination with gunpowder. Gunpowder, another Chinese invention, began to be extensively applied by the Europeans in firearms and cannon. By placing cannon on their ever larger, seaworthy ships, they created floating gun platforms. This made them vastly superior to whatever non-European people could employ in defense. On his second trip to India around Africa's Cape of Good Hope Vasco da Gama took 20 armed ships and brutally crushed Muslim traders who did not want interference in their profitable trade routes. We have already seen that this development was hardly inevitable. A century earlier the Chinese sent fleets of great junks, vastly larger than the caravels, to Indonesia, the Persian Gulf and Africa's East Coast. However, they chose not to pursue their naval expeditions any further and never exploited their invention of gunpowder to the extent the Europeans did. For them there was of course not the same incentive to discover as for the Europeans, since they were already in possession of the world's riches. Europeans on the other hand were greatly attracted to the riches of the orient, especially spices. Before 1500, spices from Asia reached Europe through intermediaries, i.e., Arab and Indian, Egyptian and then Venetian shippers. This spice trade had been very lucrative for centuries, since there was no other way for the Europeans to get their pepper, cinnamon or ginger to make their food more palatable or simply to preserve it. To capture this trade was Vasco da Gama's main purpose when he sailed from Portugal around Africa for the first time. Rather than hugging Africa's west coast, Gama used the winds to first cross the Atlantic and then swing back to the Cape of Good Hope, thus demonstrating a skill in navigation that was new to the world. From there he crossed the Indian Ocean, reaching Calicut, on the southwestern coast of India, in 1498.

It is obvious that in this story of how Europeans began to develop their colonial empires firearms and especially guns played an important role. The earliest Chinese and Arabic firearms used bamboo tubes as barrels and shot arrows. From such 'fire lances' cannon evolved among the Arabs, Turks and Europeans during the 13th century. The period 1300–1600 saw the cannon perfected, chiefly in Europe. From there, ironically, cannon were introduced into China (by Jesuits in 1520). It is unlikely that the Chinese saw no advantage in developing firearms or cannon. After all, their armed conflicts were far from over at that time. However, it is possible that Europe was more eager to have cannon than China because of its exquisite suitability to the kings of Europe's nascent states in subduing competing noblemen. The cannon was used to crush the castles from which these petty lords ran their mini-states. Soon, castles became useless and only nations could afford the advanced weaponry or fortifications that had now become necessary to wage a war. In China, feudalism had lost its footing since the Warring States period (475–221; Chapter 8). The Chinese military was shaped by the threat from the nomadic societies of Mongolia, Manchuria and Central Asia in which shock tactics and siege warfare played a much less prominent role. Also, cannon could do little against the stamped earth fortifications common in China.

Also, at that time the European climate for an entrepreneur was likely to have been better than in China because of its multiple vigorously competing states. This diversity also offered ample opportunity for a talented individual to switch allegiance and lend his services to other masters. For example, when the Turkish Sultan Mehmet II laid siege to Constantinople in 1453, he used 20-ton cannon of 26 feet long firing a 1,200 pound stone cannonball. At the time they were the latest in cannon technology and designed by the Hungarian artillery expert Urban, who had first offered his services to the Eastern Roman Empire, which could not afford him.

Perhaps the best example of the advantages the European state system had to offer the entrepreneur in those days (not related to cannon) is Columbus' proposal to use a westward route to the riches of the orient, which was initially rejected but eventually financed by the Spanish court. More sophisticated, mature societies, like ours, would never have funded him, since his calculations of the distance to China were demonstrably wrong. However, the Spanish kings decided that a little gamble could do no harm, especially since he might discover some profitable islands similar to the Canaries. Now, we would call that an irrational decision and it is a good illustration of why advanced, stable societies like ours no longer excel in discovery and invention, which are obviously irrational processes *par excellence.* After all, the presence of the American continent was unexpected. Even before Vasco da Gama landed for the first time in Calicut, Spain had begun to build its huge overseas empire in the Americas. Soon after Columbus first landed in America (still thinking it was China), Mexico, Middle and most of South America were governed by the vice-royalty of New Spain[162].

Spain would remain the dominant force in Europe during the 16th and the first part of the 17th century. At its height, its ruler in the person of Charles V of Habsburg (1500–1558), controlled not only Spain and the America's, but also most of Italy, The Netherlands and Germany (he was elected German Emperor). However, in the 17th century Europe's focus of creativity moved to the Low Countries, England and France. These countries had also begun to use the Atlantic seaboard as a basis for trade, discovery and conquest. The reason that Portugal and Spain soon fell behind was due, at least in

162 Thomas H. *Rivers of Gold: The Rise of the Spanish Empire, from Columbus to Magellan.* Random House, New York, 2003.

part, to their fanatical interpretation of the catholic religion. Spain especially made some serious mistakes by their expulsion of the Jews and their Moorish subjects after they conquered the last Arab stronghold of Granada in 1492. This meant the collapse of the Moorish system of irrigation agriculture, which could not be replaced by small farmers.

In the 17th century it was the new Dutch republic that after its successful struggle for independence from Spain would set the tone for the new age of economic growth, technology development and capitalism[163]. Not surprisingly, this began with agriculture. As mentioned above, Europe's population continued to outgrow its ability to feed itself. The Low Countries, however, were clearly the exception. We have already seen that this part of Europe was one of the richest thanks to a thriving cloth industry. Driven by a rapidly increasing population Holland, already under their Habsburg overlords, had been busy reclaiming land from the sea. They greatly accelerated these activities once they had thrown off the Habsburg yoke and established a new merchant state: the Dutch republic. With its sophisticated drainage techniques, based on windmills to pump the water out, new, highly fertile land was created. The number of cattle significantly increased due to the extensive use of fodder crops. Stalling of the cattle allowed the concentrated use of manure. Agriculture was now highly productive, especially after the introduction in the 17th century of an advanced moldboard plough, developed in China in the 3rd century BC, with ploughshares of malleable cast iron. Agriculture was transformed into a highly specialized activity producing for the market, not only those in the nearby cities but also elsewhere in Europe. This Dutch example was the basis of Europe's success at the end of the 18th century in halting the seemingly endless cycle of subsistence crises.

Increasingly rich from trade, initially in bulk goods such as timber, grain, salt and fish, but later also in high-value commodities, the Dutch republic took over the leading role of Portugal and Spain to build up its own colonial empire. Bulk-carrying shipping came to be dominated by the Dutch who succeeded the Hanseatic League. Their success in trade was even greater than in agriculture. Around 1600 they introduced their famous *fluit* ship, designed to carry maximum cargo at minimum costs. This gave them an enormous competitive advantage in commerce and very soon it was mainly in Dutch ships that the world's riches were transported from one place to

163 Israel J. *The Dutch Republic: Its Rise, Greatness, and Fall 1477–1806*. Oxford, New York, 1995.

another. They were also very active in the Asian trade, not hesitating to use military power to enforce their monopoly, against the native population or competitors. They established the Verenigde Oostindische Compagnie (VOC or United East India Company), which had its own military forces. Already at an early stage it forbade its future colony of Indonesia to trade with other countries than Holland, and it gravely exploited the population essentially until its independence in 1945 (see also below). Holland, which had its golden age in the 17th century, is often considered as the first modern economy. However, it would not take long before the other, much larger seafaring nation, England, would take over its role and establish a colonial empire the world has never seen before or since.

In contrast to Holland, England had always been a major European state, but it suffered setbacks in the 15th and 17th century due to civil war, which delayed their arrival as Europe's most successful nation. They had fought Holland's dominance on the oceans in three wars but were unsuccessful. It was only after they joined their competitor in a war against France, which led to a shared head of state between the two nations in the form of King William I, that England got its breakthrough. With South and Middle America in the hands of Spain and Portugal, the British (as well as the Dutch and the French) began to colonize North America and (later) India. All three of them also aggressively penetrated the Caribbean, Spain's backyard. Initially, all three countries established colonies in North America and India. However, a chronic lack of manpower obliged Holland soon to abandon most of its territories to the British and focus exclusively on Indonesia. Also France's possessions in North America and India would eventually end up in the hands of England, with the remainder in North America sold to the newly independent United States in 1814.

Interestingly, while all these countries were mainly interested in territories providing them with gold and silver, or high-value, warm-climate agricultural staples, such as tobacco or sugar, it would be New England in the north which did not perform these preferred colonial functions that would grow into the most powerful and wealthiest nation the world has ever seen. New England did not have precious metals or the climate to grow cash crops on large plantations worked by slaves. Instead, its free farmers enjoyed the fertility of the land in growing similar foodstuffs as in Northern Europe. These immigrants, all from Britain at this stage, also conducted fisheries, the carrying trade and shipbuilding. This made them competitors rather than a cash cow of their home country and future independence all but inevitable. Adam Smith attributed the rapid progress of the British colonies in North America to the abundance of good land and the liberty of the people to manage their own affairs. We have already seen how profitable it was to leave the peasants to their own devices.

The so-called industrial revolution of the 18th century basically originated in England as a series of social and technological innovations. There are several reasons for this pioneering role of England, but one is certainly the flow of riches from the colonies. Europe in general witnessed first the enormous increase in gold and silver from New Spain, then the profits from sugar (from the Caribbean), tobacco (from Virginia), slaves (from Africa) and spices (from the Far East). This directly and indirectly benefited most European countries. After Spain (which essentially wasted its American profits on military adventures), Holland became the first European nation to enrich itself through colonial adventures in previously unimaginable ways. This is elegantly described in Simon Schama's *The Embarrassment of Riches*[164]. After its focus on Indonesia, Holland managed to close this gigantic country off from the rest of the world, thereby creating an absolute monopoly on both imports and exports. Brought to its most extreme only in the 19th century, the Dutch introduced the so-called Culture System, which essentially turned the natives into forced labor who had to produce quotas of coffee and other products. Only Dutch manufacturers were allowed to sell on the Indonesian market. The abuses of colonialism in Dutch Indonesia are immortalized by the 19th century Dutch author Multatuli in his satirical novel, *Max Havelaar* (1860)[165]. The use of Indonesia as a cash cow basically turned Holland into a retirement state as early as 1800. Both politically and economically it transferred its great power status to England, in what is known to us as the 'glorious revolution' of 1688[166]. Like Portugal, Holland was small. No incentive for an industrial revolution there!

England would eventually get the reputation to be a benign colonial overlord and to some extent that is well-deserved. However, it is not a reason to no longer consider colonial profits as the most likely explanation for

164 Schama S. *The Embarrassment of Riches: An Interpretation of Dutch Culture in the Golden Age*. Vintage, 1987.

165 Multatuli (Edwards R, translator). *Max Havelaar: Or the Coffee Auctions of the Dutch Trading Company*. Penguin, London, 1960.

166 Jardine L. *Going Dutch: How England Plundered Holland's Glory*. HarperCollins, New York, 2008.

England's precocious industrial take-off. Unfortunately, possibly as a reaction to the guilt-driven 1970s, some prominent authors seriously proposed cultural qualities instead of colonial enrichment as the main explanation of why European — especially English — societies experienced a period of explosive growth when the rest of the world did not. For example, according to David Landes, in his 1998 book *The Wealth and Poverty of Nations*[167], the 'European Miracle' was driven by factors as the protestant work ethic (originally proposed by Max Weber), absence of 'Oriental Despotism', strict adherence to the rule of law, property rights, thrift and entrepreneurship and honesty (trust lowers transaction costs). In a recent book, Gregory Clark goes so far as proposing that the Industrial Revolution was the gradual but inevitable result of a kind of natural selection, in which economically successful individuals transmitted to their descendants, culturally and perhaps genetically, such productive traits as thrift and devotion to hard work[168].

Britain is considered by such eminent authors as Paul Johnson, Niall Ferguson and the above mentioned David Landes as at worst an enlightened despot. They seem to argue that its colonies were a hindrance rather than an asset, because of the high costs of maintaining order and assisting the colony in developing itself. And it is true that in the later 19th and early 20th century Britain did show ample signs of a benevolent paternalism, often at the expense of the taxpayers at home. But its goodness can be exaggerated too. In his book *The Birth of the Modern*, Paul Johnson, for example, explains how British administrators in India set about improving the lot of ordinary Indians and contrasts this with the native rulers, who he describes as "men who were indifferent to their subjects' welfare, lazy and avaricious at best, and actively cruel and aggressive at worst". The British by contrast are described as "going through a progressive, peaceful revolution in which governmental corruption was being stamped out, criminal law reformed and almost every aspect of daily life altered for the better"[169]. Yes, and all of that because the British had gone through generations of genetic or cultural enrichment, gearing their people towards entrepreneurship, benevolence,

167 Landes D. *The Wealth and Poverty of Nations: Why Some Are So Rich and Some So Poor*. Norton, New York, 1998.

168 Clark G. *A Farewell to Alms: A Brief Economic History of the World*. Princeton University Press, 2007.

169 Johnson P. *The Birth of the Modern: World Society, 1815–1830*, p. 794. HarperCollins, New York, 1991.

trustworthiness and efficiency! Does this sound too good to be true? Then probably it is not true, or at least not exactly true.

Without disputing Johnson's outstanding story of how Europe entered modernity in the 19th century, thereby changing for the better, it is important to realize that it had not always been like that. Indeed, the industrial revolution at the end of the 18th and the beginning of the 19th century was entirely driven by events taking place in a not so benevolent time. The typical British mindset so praised by Landes, Clark, Johnson and others was not typical for Britain or Europe at all, but simply a consequence of the competitive state system so characteristic for Europe since the Middle Ages. We have seen similar mindsets at work in early China and the Hellenistic states of the Mediterranean basin. It was this competitive, hardnosed state of mind that facilitated Holland's and England's ascendance to become the leading nations in Europe. And of course, as already pointed out, Europe had all the advantages of a late-comer; it cultivated the right mindset when a lot had already been invented by others and further optimized by their ancestors from the Middle Ages.

In the 17th and 18th century, British citizens (not necessarily the British State) became rich from the sugar industry in the Caribbean, from the slave trade, and most of all from the conquest of India. Here, they profited enormously from the power struggle that had resulted from the disintegration of the Mughal Empire. There were not many signs of British benevolence during these early stages. Indeed, in those days the British were as corrupt and morally flawed as they found the Indian native rulers. They just showed more discipline, persistence, steadiness and ferocity than the Indians and were also better trained with better weapons. As individuals, they were determined to get what they wanted.

The early British adventurers-turned-statesmen, such as Robert Clive (1725–1774) and Warren Hastings (1732–1818) were in it first for the money, their own private money, that is, and then for the aggrandizement of their nation. Their fearless (we would now call it irresponsible) behavior, supported by a persistence of mind that by now has become incomprehensible in our own society, brought them power and riches. As it is now to us, such a state of mind had become alien to the sophisticated, civilized Indian society, producing its riches since ages under the stable direction of the Mughal overlords. Therefore, arguments of moral superiority of 18th century Europeans, whether British, French or Dutch, are not only insulting to non-Europeans, but also without factual basis.

Just to make things clear, Niall Ferguson, also a great proponent of the benign British overlordship, reminds us that "this was how the British Empire began: in a maelstrom of seaborne violence and theft"[170]. And true enough, after his victory at Plassey (1757), Clive's cash reward was the equivalent of $140 million. Since this was far from the only one of his many rich pickings from India, Clive, who started out as a poor kid when he began working for the East India Company, exemplified the 'nabob', a corruption of the Indian title of *nawab*. When he returned to England, he was rich and could buy himself real estate and a seat in parliament. So much for the absence of corruption!

Clive was far from alone. British adventurers amassed enormous wealth from India. Custom dictated that soldiers were given prize money collected from the spoils taken during a campaign. The largest beneficiaries of the prize system were the senior commanders. At this early stage in British colonialism there is therefore no evidence of moral superiority vis-à-vis the native rulers. It was only later that the moral improvements of the British described by Paul Johnson came into play (possibly as a consequence of the economic and social advancements that followed the industrial revolution) in the sense that in the age of the Victorians, it was considered a duty to uplift and "improve" the subjects in the colonies. It was then that the stereotype came into vogue that the people of India were timid and servile (comparable to the picture of ridiculous Italians or lazy Frenchmen, who "should do better"). Nonetheless, it should be noted that while British subjects saw enormous improvements in their living standard, the colonial empire was less beneficial to the mass of British citizens than it was to the small minority that was invested overseas. Indeed, while the benefits of overseas investments were not enjoyed by the majority of the people, they were the ones to foot the bill for the costs of governing the empire and for its defense.

Of course, Europe and especially Britain enjoyed cultural and political factors conducive for an industrial take-off. But they were not the first to go through that stage. Other societies in history had created similar opportunities for themselves. The fact that in Europe it was England that was the first to take off as an industrial power had nothing to do with a unique

170 Ferguson N. *Empire: The Rise and Demise of the British World Order and the Lessons for Global Power*. Penguin, New York, 2002; emphasis mine.

mindset. Indeed, as stressed by other authors, the random elements in technological progress are important and often decisive. As usual, the critical factor was an increase in agricultural productivity. Agriculture was always more productive in Northern Europe than in the Mediterranean Basin, once the tools were in place to work the heavy but fertile soil. In 17th century England, great strides were made in agricultural improvements. Marshes were drained (by a Dutch engineer) to gain more land for farming. The enclosure laws permitted lands that had been held in common by tenant farmers to be enclosed into large, private farms that could be more efficiently worked by a much smaller labor force. The use of seed drills (invented in China) allowed farmers to sow seeds in well-spaced rows at specific depths, which boosted crop yields because more seeds could take root. New methods of crop rotation that improved upon older methods, such as the medieval three-field system discussed earlier, played a major role.

England also had the natural resources that permitted an industrial revolution, such as water power and coal to fuel the new machines, iron to construct machines, rivers for inland transportation and good natural harbors. An important factor certainly is the gradual change in values, with the values of the mercantile and capitalist classes slowly becoming the norm. Indeed, this was the time of Adam Smith's *The Wealth of Nations*, in which he proposed that the only legitimate goal of national government and human activity is the steady increase in the overall wealth of the nation.

Finally, it is often forgotten that Britain did not become the world's leading power because of its adoption of a parliamentary system and the rule of law, but because of its capability to make war. The intense competition with other European powers, most notably France, had turned England into a lean and mean fighting machine. A major factor in England's superior performance in making war was the establishment in 1694 of the Bank of England, built on the modern system of privately financed public debt pioneered by the Dutch. This gave them the opportunity to raise funds quickly and gain the upper hand in virtually all major European conflicts until modern days.

The key factor in England's take-off as the first industrial power, however, remained the colonial opportunities offered to England and other European nations, opportunities they grabbed but were denied to other societies of equal mindset that went ahead of them. Colonial opportunities brought the capital and capital is necessary for growth. It was their good fortune to accumulate all that capital at the right time and in the right geographical place. Yes, England had a labor shortage, and it had the good fortune to sit on coal mines, and it had a parliament-dominated government, which was inclined to give commerce free rein, but this would not have been enough to guarantee risky investments in new technology. While conditions elsewhere in the 18th century, in India or even in most European states, were not conducive to an industrial revolution, England could not have done it without the inflow of private capital from its colonies, without the raw materials from these same colonies and without the large, captive market of a colonial empire provided at a somewhat later stage. Once the industrial revolution was a fact and modern Europe well on its way, the impression was easily made that inherent characteristics, such as incorruptibility or thrift, were the reasons for success rather than colonial profits.

Hence, Europeans were not inherently different from Asians or Africans or Native Americans. It really was a combination of natural advantages and sheer luck that first allowed Europe to develop into aggressively competing states exploiting their edge in agriculture, fully taking advantage of their window on the world through the Atlantic seaboard and becoming a global power. Then, the wealth derived from its colonial adventures fueled the industrial revolution and established the European global Empire of which we are now all a part. It is certainly possible that others before them had some possibility to do the same thing. For example, Rome at the end of the republic was merely a robber state accumulating such tremendous wealth that some of its beneficiaries did establish production centers almost on an industrial basis. But the institution of slavery as well as the consolidation of power by the emperors prevented actual take off. It has been argued that also in Ming China fabulous, concentrated wealth could have led to industrial take-off, but this is debated[171]. However, in contrast to the situation in Europe, in China many of the great industrial centers were in government hands and in the absence of strong competing states after the Ming, the Empire may have been less inclined to invest in increased productivity.

From the 15th century onwards Europe's colonies yielded a continuous stream of riches from the mines and the plantations, worked by slaves, which made its owners in Europe extremely wealthy. Improvements in ships and shipping as well as the great increase in total ship tonnage promoted divi-

171 Mote FW. *Imperial China 900–1800*, p. 765. Harvard University Press, Cambridge, 1999.

sion of labor since food could now be imported on a large-scale, freeing up many hands for manufacturing. Like in China, great improvements in agriculture and the adoption of new food crops first cultivated in America, such as maize and potatoes with their high average yield in calories per hectare, contributed to the surge in population after 1500. Moreover, especially in North America, the Europeans had now access to extensive, fertile new lands, largely emptied of native people by the exported diseases, such as small pox. This provided them with the opportunity to shed their surplus population.

Most importantly, Europe's colonialist nations were able to use their military power to obtain the necessary capital, raw materials and markets to begin manufacturing at a scale that had never been seen before. European trade and manufacture stretched to every continent and this vast increase in the market for European goods in part drove the conversion to an industrial, manufacturing economy. Other nations could not initially do the same thing because of the monopolistic control that the Europeans exerted over the global economy. World trade was there to make Europeans wealthy, not to enrich the colonies or non-Western countries. The best example is the cotton industry.

The industrial revolution began with cotton. Cotton was originally domesticated in India, as we have seen, and by the early 17th-century India had advanced pre-industrial cotton manufacturing. Imported Indian cotton fabrics became very popular in Britain, competing with the silk and wool weavers. Once Britain had conquered a substantial part of India's territory, it was in a position to put a ban on Indian textiles and provide a stimulus to have India's raw cotton being shipped to the British cotton mills in Lancashire. (England's reputation for free trade stems from a later time.) India was now reduced to a provider of the raw material for the mills of Lancashire and a market for the finished product.

The increasing demand for cotton fabric stimulated manufacturing and led to a succession of inventions, from John Kay's flying shuttle for weaving in 1733, James Hargreaves' spinning jenny in 1767, which enabled a spinner to spin several cotton threads simultaneously, Richard Arkwright's water-powered spinning machine in 1768, which combined rollers and spindles for spinning, Samuel Crompton's spinning mule in 1769, a cross between the jenny and the Arkwright spinning machine, Edmund Cartwright's power loom, a mechanical weaving machine, and Eli Whitney's mechanical gin (engine) for separating cotton seeds from fibers. In time, this led to similar innovations in the wool and linen industries. The textile revolution was basically completed before the end of the 19th century.

It is the mechanization of the cotton industry that started the industrial revolution. In turn, its profits financed the railroads, which then facilitated the expansion of metallurgy, which became crucially important after the invention of the steam engine by Papin, Savery, Newcomen and Watt. Watt's engine soon became the dominant design for all modern steam engines. Steam engines powered all early locomotives, steam boats and factories and are often considered as the foundation of the Industrial Revolution. The earliest modern steam locomotive was Stephenson's Rocket, built by George Stephenson in 1829.

These developments greatly increased demand for iron, cast iron and especially steel. It was not until the mid-nineteenth century that Henry Bessemer learned, what the Chinese already knew much earlier, how to make steel in vast quantities and at prices that could compete with wrought iron. As discussed in Chapter 8, this process involves the introduction of air into the fluid metal to remove carbon from molten pig iron. Like in China, the Europeans began to use coke instead of charcoal for smelting iron ore. In Europe, this fuel switch was important because of the increasing scarcity of wood relative to coal, which was available in large quantities in England, the US and continental Europe. The technical improvements, the new fuel and the streamlining of operations made iron and steel for the first time available to Europeans at a very large scale.

The introduction of steam power (fuelled primarily by coal) and powered machinery (mainly in textile manufacturing) resulted in dramatic increases in production capacity. The development of all-metal machine tools in the first two decades of the 19th century facilitated the manufacture of more production machines for manufacturing in other industries. Scientists, initially merely theoretically oriented, now began to interact with the artisans and contribute to technology development. This is especially true for chemistry. The period from 1850 to 1914 saw the development of the chemical industry and the professionalization of chemistry. Organic chemicals, often synthesized in the laboratory, were being more widely used for pharmaceuticals, dyes, explosives, and fertilizers. Chemistry combined technology and science to become one of the most important industries in the period, which is sometimes called the Second Industrial Revolution.

Later this would apply to other sciences as well and nowadays many people hardly see the difference between science and technology.

After the gradually perfected steam engine, the next major innovation that fueled the modern era was electricity. Still considered a curiosity in the 18th century, Michael Faraday discovered electromagnetic induction, the generation of an electric current by revolving a magnet inside a coil of wire. This led Samuel Morse to develop the electric telegraph in 1838, which would quickly replace the semaphore telegraph, based on towers conveying information by means of visual signals (using pivoting blades or flags) and itself only introduced a few decades earlier. In 1873, Benoît Fourneyron developed a turbine driven by water for electric power generation. The steam turbine, developed in the 1880s could do the same but was now burning coal, which made it independent of streams and rivers. Coal-burning electrical power stations are still our main producers of electricity today with the serious environmental consequences we are now so familiar with. In 1891 a 175-km long transmission line was built in Germany: the forerunner of our modern power transmission system called the 'grid'.

Electricity soon proved extremely convenient. Electric arc lamps were already replacing gas light in the 1850s (introduced half a century earlier, thereby making cities a whole lot safer). However, the main breakthrough was the invention of the incandescent electric lamp between 1878 and 1880, simultaneously in England and the United States. The device has been with us ever since and it is only now that its notorious inefficiency is being recognized and a replacement sought. An electric tram (now, we would call it 'light rail') was invented in Germany in 1879, with subsequent applications of electricity in industry and as household appliances.

A key invention which we now desperately seek to replace was the internal combustion engine to drive motorcars and airplanes. Already in the beginning of the 19th century a system of well-designed roads had greatly improved transportation by coach, and at the end of that century an extensive railway system was in place in England, the US and most of continental Europe. Coaches were slow, however, and trains had the same disadvantages as they have nowadays: they sometimes do not go to places where you want to go or their schedule does not fit yours. The internal combustion engine running on gasoline proved much more suitable to drive relatively small, private vehicles than the steam engine. Hence, around the turn of the century we see the rise of the automobile industry. Shortly thereafter, an internal combustion engine was used to drive the first airplane in 1903. Our present-day communication palette was virtually completed by the invention of the telephone by Alexander Graham Bell in 1876 and wireless telegraphy (or radio) in 1892 by Nikola Tesla.

A major development in manufacturing was not an actual invention but the routine introduction of interchangeable parts to make complex products. Pioneered by the Dutch on their shipyards during the 17th century, this was now quickly becoming a standard feature and nowadays manufacturing without interchangeable parts cannot be imagined.

There can be no doubt that it was the huge surplus generated by its overseas possessions that drove Europe's industrial revolution. It is not a coincidence that Britain, so successful in this respect, was the first to take off. Hence, in a real sense, Europe's unparalleled wealth-generating machine was realized over the backs of non-Europeans. However, Europe's dominant class (now no longer the great landowners, but the great industrialists) did not treat its own workers any better than their colonial subjects. The new factory system used its workers in a systematic fashion. Often including women and children, they had to work long hours for very low wages under embarrassingly bad conditions. Anybody who ever read the works of Charles Dickens knows what it meant to be poor in 19th century England. The new system also put traditional workers like the handloom weavers out of business. It was ruthless capitalism but it was profitable.

We have already noticed that increased wealth and stronger governments lead to increased humanitarianism. This was true for the Roman Empire and for ancient China. It was also true for Europe. In spite of its poor moral record of colonialism and the cruel treatment by its industrialists and landowners of their own workers, Europe did find an effective social and humanitarian approach that has now gained global acceptance. Also this new state of mind was pioneered by the British. They took the initiative to abandon the slave trade (in 1807) and, in contrast to the Dutch (see above), they also reigned in during the 19th century the use of their colonial subordinates merely as coerced labor. The best that came out of this, however, was a slight increase in Indian life expectancy and the average Indian never came to share in the enormous increase in the British per capita gross domestic product.

In Britain, as elsewhere, a collection of social provisions gradually became enshrined in law. This has been so successful that the original socialists would undoubtedly consider modern Europe or even the United States a socialist paradise. Western civilization crafted freedom as its distinctive leitmotiv, which became the norm and practice for private life and government. This was enshrined in 1948 in the Universal Declaration of Human Rights at the United Nations. Since then there have been setbacks with some nations contesting elements of this culture of freedom, but overall it has been successful. Since 1975 the share of free countries increased from 25 to 45%[172].

It is important to see this development for what it was: not an obvious, universally valid and principled choice of a select culture, but a natural consequence of increased wealth. We have already seen that in the history of humankind dictatorships, not democracies have been the norm. Social care and freedom have origins closely associated with Europe's industrialization. While individual reformers (usually independently wealthy) undoubtedly played a role in the moral outrage that contributed to social improvements, such as the abolition of slavery, shortening of the working day, increasing of wages, improving of working conditions and extending the franchise, such advances were mainly the result of a determined struggle by large groups of people realizing their own power in the relatively open and increasingly wealthy European society. Organized labor in the form of trade unions and political parties waged campaigns to raise the common standard of living and enforce equal participation in the state's representative institutions. Threatened by military force, often arrested and imprisoned for longer periods of time, these people and their leaders were able to eventually pull it off. Indeed, victory was so complete that in our own days there is no longer a working class and socialism has outlived itself. Soon social provisions began to be adopted on a global scale, even in the notoriously conservative United States. In the 20th century we see the virtual completion of a society in which a decent standard of living and freedom are considered basic rights.

It is not unreasonable to state, as Paul Johnson does in his aforementioned book *The Birth of the Modern* that the matrix of the modern world was formed in the years between 1815 and 1830. Already at the end of the 19th century, the picture of the world as we now know it was virtually complete, at least in rudimentary form, in Europe and the United States. At the beginning of the 20th century, European supremacy was generally acknowledged and its technology had begun to be installed everywhere on the globe. Also

172 http://www.freedomhouse.org

many social and economic features of European society, understood to be an essential component of its success, were adopted everywhere. The best example is of course Japan, which was opening the doors to trade with the United States in 1854 under pressure of the naval officer, Commodore Matthew Perry, after two centuries of isolation except for some isolated contacts with a few Dutch and Chinese traders. This was followed by the Meiji restoration in 1868 that led to the restoration of power from the shogunate to the emperor and the series of revolutionary changes that coincided with the Meiji emperor's reign (1868–1912), the downfall of Japanese feudalism and the forging of a new and modern state. Its military power was demonstrated in the first Sino-Japanese War (1894–95) and the Russo–Japanese War (1904–5). Other countries would follow and now, more than a century later, confidence in European technology is so strong that few would seriously entertain the premise of this book, i.e., that technological progress has tanked with novel, groundbreaking inventions and innovations lacking already for decades.

The 20th century saw more macro-inventions and many innovations. Earlier accomplishments were optimized and their application greatly expanded. Automobiles began to be mass produced with the moving assembly line of Henry Ford as early as 1913. The extensive road systems already in place halfway the 19th century were now greatly expanded and made suitable for our current massive use of the motorcar. This reached a high point with the National System of Interstate and Defense Highways (about 41,000 miles of roads) in the US in 1956, signed into legislation by President Eisenhower. As the name suggests, this system was partly designed to move military equipment and personnel efficiently. As described in Chapter 3 of Part One, the system was copied from Germany's autobahn network, built before the Second World War with the same purpose in mind. Nowadays virtually all of these networks are freeways, access-controlled, divided highways of at least two lanes each direction. Almost all countries now have freeways, illustrating the dramatic success of the motorcar as a consumer product.

Commercial aviation seriously took off in the 1930s, with its first transatlantic service available on the eve of World War II. With the introduction of the modern jet plane in the 1950s in England, France, the former Soviet Union and the US, its development was essentially completed. An invention that can be assigned to the 20th century and is often considered as one of the most important breakthroughs ever made is the personal computer. It was the personal computer that led to the internet and the eventual convergence of a broad range of informational tools and services. The impact of these macro-inventions in tying the world together into a global playing field for individuals to generate wealth is vividly illustrated by Thomas Friedman in his book *The World Is Flat*[173]. This development was greatly facilitated by the end of the cold war in 1989 (with the fall of the Berlin wall), and the inclusion of the states of Eastern Europe, as well as Russia and China, in the world economic system. This enormous increase in political stability heralded increased wealth and individual freedom as the gold standards for successful nations, but simultaneously, as now appears to be the case for all successful societies, a decline in the rate of innovation. Thus, instead of providing the perfect environment to foster further and much more dramatic progress than witnessed hitherto, this new goldilocks period appears to be giving us technological slowdown.

In Part One we saw that, from whichever way you look at it, technological progress in the 21st century is on the decline. Surprisingly, this seemed to find its roots in our very success of building up a stable society, where most people can prosper and live a good life, free from fear of having their life and property taken from them. Such a society is sustainable over the long term because it is based on the dependable foundations of the rule of law, allowing for orderly, peaceful change rather than continuous upheaval. In short, it is a society that as described in Part Two has been emerging multiple times in human history, from the original hunter-gatherer societies, through the early Neolithic village cultures to the great empires and our own global economic system of states. Technological progress stalls not during wars and other upheavals, but precisely at our finest moment. We seem to be close reaching that stage. But there is a difference with previous cycles.

The main difference between then and now is that this time there are no longer external threats, such as foreign invasions, to ultimately set in motion another cycle of innovation, success and stasis. Neither the Romans nor the Chinese experienced a gradual decline due to internal rot. Internal rot theories are fantasies mostly coming from Western authors. They are fantasies based on the desire to see more in their own success than a fortuitous combination of geographic factors and historical events. Europeans never

173 Friedman TL. *The World Is Flat: A Brief History of the Twenty-first Century*. Farrar, Straus and Giroux, New York, 2005.

developed peculiar skills to make them more successful than Asians, Africans or Romans. They were not superior to the Chinese in the 15th century. And Europeans did not invent mass manufacture and industrialization; China and India did that, exporting goods on a mass scale to every corner of the world, similar to what they are doing now.

Therefore, the single major difference between the ancient, successful civilizations and us is the absence of external threats to break through our period of stasis once this has settled in. Now, 'us' means all of us. Us is Europe and us is North America. Us is also China and India, and Japan, the nations of the African continent and South America. Us is now the whole planet. While the Romans always had to reckon with barbarian tribes on their northern borders and with the Persian Empire, China had to fight off inner Asian tribes, and many an Indian civilization expired by aggression from the North, our current world society appears to be safe from such threats. Part of my thesis is that it is really external threats and nothing more that can end a successful civilization. Does this mean that we do not need to worry about the consequences of our own technology slowdown? What will happen to us when new macro-inventions will no longer appear? I will try to give the answer to that question in the Epilogue.

Epilogue — Shattered Dreams: The Future of Innovation

This book is an attempt to understand why a society at some point in time experiences a rapid accumulation of major inventions, leading to a flowering and highly successful civilization, only to see further technological progress stall when conditions for further expansion have never been better. As described in Part One of this book, our own society is currently entering such a period of technological stasis, which is associated with increased peacefulness, openness, education and general well-being, all as a consequence of increasingly good governance and international cooperation. We favor stability, reasonableness and compromise.

In Part Two we have seen that technology ups and downs are recurrent events in history. Technological progress thrives in times of upheaval and vigorous competition among states and systems of governance. New inventions that help secure dominance are highly desirable, with turbulence and uncertainty making their adoption relatively easy. Qin Shi Huang, who unified China in 221 BC (the one of the terracotta warriors) had little patience with the status quo. He vigorously adopted major innovations, varying from military conscription and the modern horse collar to the construction of the Grand Canal, connecting China's many rivers into a gigantic transportation network.

However, the very success of such states carries the seeds of their own demise. Successful, highly stable states have great difficulty in adopting new inventions. Novelties infringe on established patterns of how things are done. They experience resistance, from the public, government and industry, due to such seemingly rational considerations as costs, safety and a lack of obvious long-term advantage (as if we ever predicted the astounding success of the motor car or airplane!). In the early days, when successful societies are born, novelty is not scrutinized at so many levels as it is later, when society is mature and the structures are in place to effectively resist it. Once states and ideas are reconciled and society stabilizes, new inventions are no longer easy to implement.

The reason that highly stable, successful societies keep their population from innovating by resisting the adoption of new, breakthrough inventions is a direct consequence of their very success. It is not only the ruling oligarchy and higher strata who are generally well-satisfied with the existing situation, but also most of their less well-off subjects fear that change would make them worse off. We see this not only in our own planet-wide economic empire but also in the wealthy state systems that arose at different places in the times preceding us, such as the Roman and Chinese empires. From contemporary writers we know that both the Roman and Chinese citizenry were convinced of the superiority of their states and the need to maintain the status quo. The average Roman citizen did not know that his or her Chinese counterpart was significantly better off, and neither could have a clue how much society would have improved by the end of the second millennium. Indeed, in their own time they were looked at with envy by their barbarian neighbors, who tried to emulate them. In spite of their military depredations, many of these people eventually wanted nothing more than to be part of the Roman or Chinese society and live under conditions that were undoubtedly the best imaginable for those days. Today we see something similar, perhaps, with emigration from poor countries into the US and Europe.

Importantly, stalled technological progress does not coincide with decline. While nowadays most books on the topic associate the Roman and Chinese twilight of empire with technological inferiority, contemporaries thought very different about the matter. Loss of inventiveness is not the same as loss of technology and has certainly not been a cause of the collapse of major states. Although there are some examples of 'lost technology', most notably the loss of the Roman method of making cement, the ancient Roman or Chinese civilizations remained perfectly capable of operating iron works, water mills or spinning wheels.

In the absence of new breakthrough inventions, a society can still maintain its existing repertoire of tools and continue to make improvements. In China and the Roman Empire technological stasis coincided with the heights of their power and a period in history often considered the happiest. And so it is with our own society. We may never set foot on Martian soil or routinely fly from New York to Tokyo in a few hours, but we have organized our society pretty well. As discussed in Chapter 6, conditions of humankind almost everywhere on this planet have improved dramatically.

Stable and non-innovative societies are not inflexible and can respond effectively to an altered situation. The Roman Empire in the third century may have reached technological stasis, but it was perfectly able to make the innovative changes necessary when a new major threat in the form of an invigorated Persian Empire rose from its ashes. Rome suddenly had to deal with wars on two faraway fronts. The subsequent re-structuring of its fiscal administration and changes in its military organization were prudent, led to the restoration of peace and stability, and did not cause upheaval. These innovative measures guaranteed a mostly peaceful continuation of the empire for another 200 years. In China, the agricultural innovations under the Qing in the 18th century were highly effective as the continuous rise in population shows. China may have been a technological fossil in those days, but it clearly could implement changes when necessary.

The situation is somewhat different in our own times. Our society is not threatened with invasion by barbarian steppe people who are militarily more powerful because of their masterly control of the horse. Some would argue that international terrorism is today the equivalent of the barbarian threat in those days. However, such a comparison overlooks the fact that while today's wealthier nations, as democracies, may often be considered benign, their law enforcement and military machineries are extensive, effective and persistent. In the long run they will always prevail.

So, if we have neither external nor internal enemies bent on our destruction, what stands in our way? Are there no consequences of a technology decline other than that we will never meet our 'singularity' or enter a 'Jetsons' age?

Consequences of a decline in technological progress

A decline in technological progress does not mean that we are slowly forgetting how to manufacture, operate or maintain the advanced tools that have become so characteristic for our society. We will undoubtedly continue to make cars, airplanes, washing machines and cell phones until far in the foreseeable future. While there may no longer be many breakthrough inventions, we will continue to make improvements in current technology. Examples are numerous. In the coal, oil and gas industry improvements will keep exhaustion of fossil fuels at bay for a long time to come. Improvements in alternative sources of energy, such as biofuels, solar and wind, together with improvements in the electric grid can increasingly allow us to reduce pollution and reduce our reliance on limited fossil fuels. Improvements in transportation will not produce dramatic changes but will continue to increase safety and convenience. Developments in medicine may not eradicate major killer diseases as in the past, but they will nevertheless gradually improve care. What we will not see are dramatic changes that alter the way we live or work. There may never be nuclear fusion plants, hypersonic passenger airplanes, computers to communicate with or cheap space travel. What then are the consequences of stalled technological progress?

One major consequence of a decline in technological progress is obvious to everyone, and that is the cessation of further increases in wealth. In the introduction to his book *The Lever of Riches*, Joel Mokyr explains that technological progress has provided society with a 'free lunch', defined as "an increase in output that is not commensurate with the increase in effort and cost necessary to bring it about"[174]. In other words, technology greatly increases per capita wealth. Hence, a deceleration of the rate of innovation leads to stalling or even declining per capita growth.

According to most economists long-term economic growth, i.e., the increase in wealth generated by being more productive, is mainly due to technological progress[175]. This was first clearly articulated by the American economist Robert Solow in 1956[176]. The key factor here is productivity, which basically stands for the efficiency of the labor force to make the products we need and crave. It is through productivity gains that we are able to increase our living standards. While productivity can rise temporarily, for example, when

174 Mokyr J. *The Lever of Riches: Technological Creativity and Economic Progress*, p. 3. Oxford, New York, 1990.

175 Mokyr J. "Chapter 17: Long-term economic growth and the history of technology". In: Aghion P., Durlauf SN (editors), *Handbook of economic growth*, Volume 1B, Elsevier, 2005.

176 Solow RM, "A Contribution to the Theory of Economic Growth." *Quarterly Journal of Economics (The MIT Press)* 70 (1): 65–94. 1956. doi:10.2307/1884513. http://jstor.org/stable/1884513.

workers work harder during periods of high demand, permanent increases in productivity are due to improved or new technology. Therefore, one would assume that when expenditures on research and development are continuously increasing, as they did in advanced countries during the period after the Second World War, productivity and economic growth would follow suit. Paradoxically, both have declined since the 1950s (Figure 1).

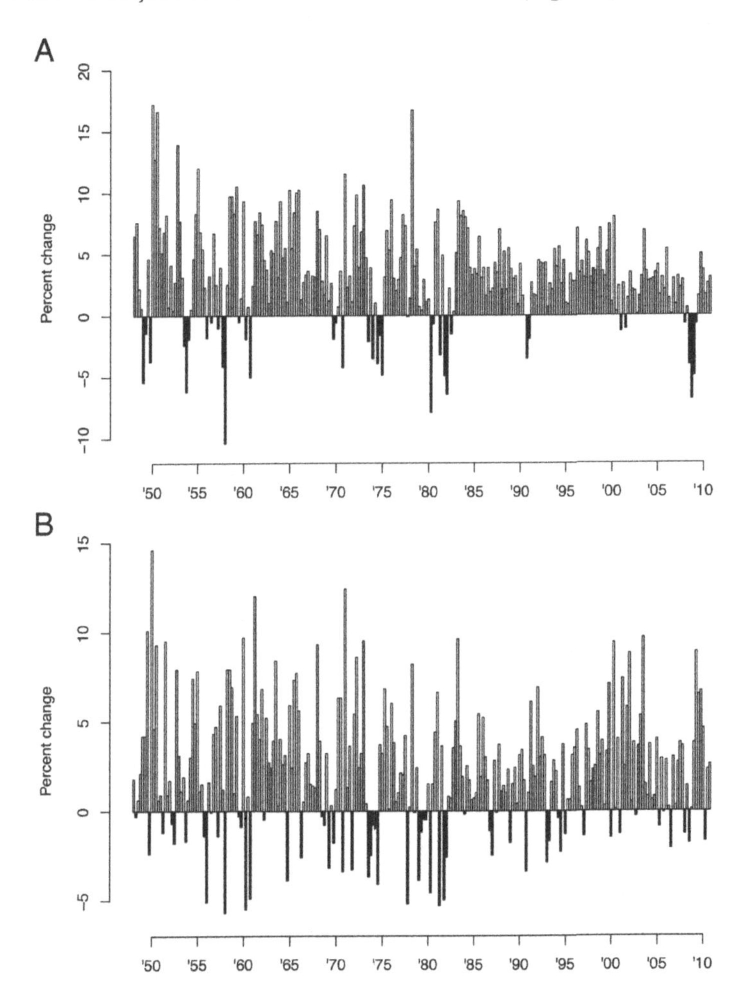

Fig. 1. Economic growth (A) and productivity growth (B) since 1950. Source: US Bureau of Economic Analysis and US Bureau of Labor Statistics.

It is important to look at productivity and economic growth in the long run because both are subject to fluctuations creating booms and busts. For example, while the US economy experienced a productivity slowdown from the mid-1950s to the mid-1990s productivity growth then rebounded and started to follow an upward trend (Figure 1). This is most likely due to the delayed effects of increased investment in information technology, which is the latest area of technology development where progress was translated into practical applications, mainly in retail, wholesale and financial services. In Europe, on the other hand, productivity growth has experienced a decline since the mid-1990s, possibly because of less use of information and communication technology[177].

Some economists explain the downward trends in productivity and economic growth by proposing that firms have begun to make equivalent innovations rather than non-equivalent innovations[178]. This means that rather than trying to come up with different products or processes they try to find more applications for the same product or process. For example, rather than developing revolutionary new drugs the biopharmaceutical company Genentech tried to market its existing colon cancer drug, avastin, also for breast and lung cancer[179]. This is called product (or process) differentiation, which has only limited effects on economic growth. What this tells us is that people started to invest in less riskier projects over the last decades, which explains the reduction in macro-inventions since about 1970. I already discussed this development in Part One, mainly in Chapter 6. A practical consequence of this is that we should not expect automatic increases in our paychecks anymore or a further expansion of luxuries within our reach.

And, indeed, when we look at the average family income in the most developed parts of the world, i.e., Europe and the US, we see that this is no longer dramatically increasing after 1970 as it did earlier. At the end of the most recent economic expansion, from 1995 to 2007, the median family income in the US of about $60,500 was less than in 2000[180]. No one except the very rich escaped this decline in family income. While this was not the first time that the lowest part of the income scale did not profit from an

177 Gomez-Salvador A., Musso A., Stocker M., Turunen J., European Central Bank, Occasional Paper Series NO. 53 / October 2006, "Labour Productivity Developments in the Euro Area", http:/www.ecb.int

178 De Loo,Ivo & Soete,Luc, 1999. "The Impact of Technology on Economic Growth: Some New Ideas and Empirical Considerations," Research Memoranda 017, Maastricht : MERIT, Maastricht Economic Research Institute on Innovation and Technology.

179 Avastin's therapeutic value is already very small, even for colon cancer, in spite of its price tag of $100,000 annually.

180 Leonhardt D. "For many, a boom that wasn't". *The New York Times*, April 9, 2008.

economic expansion, it is probably the first time that the middle class lost ground. While real median family income more than doubled from the late 1940s to the late 1970s, it has risen less than 25% in the three decades since then.

Hence, for the first time since the 18th century, wealth for the average citizen in the US, the most powerful and innovative economy, no longer grows. This does not mean that growth has stalled for everyone overnight. There are still increases in wealth for large segments of the world population. Within the US, for example, some sectors in the economy, most notably the financial sector, have seen significant growth lately, even after the financial crisis of 2008[181]. Also developing nations will grow for quite a while, depending on the quality of their governments, simply due to the enormous catch-up they have to do. For example, a typical farm in sub-Saharan Africa produces its food grains about 1000-fold less efficiently than its counterpart in Iowa. This leaves significant room for productivity increases, simply by infusing existing technology into African farms. However, while this would lead to higher average wealth and a more equal distribution of the benefits of existing technology, it would provide no novel driver for renewed productivity increases in nations at the summit of the innovation ladder.

Reduced wealth can also be camouflaged by social measures. Europe is a clear example of how increased sharing ('spread the wealth around') can easily give the impression of an increasingly wealthy population while the absolute numbers speak otherwise. Stalling technological progress can also have an effect on commodity prices. For example, food prices are no longer decreasing as they used to do. In fact, they have gone up quite dramatically lately, due to the rapid increase in people all over the world buying ever-greater quantities of food, by itself of course a very good thing and another sign of a successful society. In the absence of great technological breakthroughs, food will never again dramatically decrease in price.

The rapid increase in luxury goods within reach of all but the poorest citizens has already halted. In the absence of new, much more powerful airplanes, airline tickets will no longer continue to become cheaper and we will no longer enjoy more value for our money in the form of increased comfort and speed. Driving a car will probably become routine for most citizens

181 Some would argue that the innovative practices that gave rise to US productivity growth at the end of the 1990s were in fact the main cause of the financial crisis.

of China and India too, but it will not become much cheaper. Indeed, while peak oil may be a long way off, prices will go up and global warming will also necessitate a switch to electric cars. Electric engines are simply not as efficient as their internal combustion predecessors, a fact that translates into higher price tags.

Even our most beloved area of technology development will not be spared. Apple's iPhone or the Facebook social networking service are both based on technology developed several decades ago in times when the risk of the new could still attract investment dollars. In the absence of well-funded universities and great corporate research labs, all free to do the basic science they like rather than the 'applied' and 'industry relevant' activities dictated by the rational voices emanating from corporate headquarters, it is simply unlikely we will again see the dramatic technological breakthroughs that in the past led to the generation of such enormous wealth increases. However, is this necessarily bad and will it lead to collapse?

Thus far, stable societies always collapsed under the influence of disruptive external forces. In Chapters 8 and 9, I described how the mighty Roman and Chinese Empires, still at the top of their games, fell as a consequence of barbarian invasions. Our society, however, is the first truly global society that has managed to absorb all its rivals into one shared economy. A nuclear Iran and North Korea, or Moslem terrorism, are in reality not more than minor irritations. The bulk of their populations will be bedazzled in their turn by the lure of relative prosperity. It also seems unlikely that today's engineers will forget how to maintain and repair the tools they currently operate, as Asimov described so vividly in his *Foundation Trilogy*. As in all successful societies, few of us have disaster on our radar screens. But, unfortunately, disaster would provide the surest way to get technological progress back on track.

A RETURN TO INSTABILITY?

The eventual outcome of our technology slowdown is obviously unpredictable. Like the Romans and Chinese, who were unable to anticipate the barbarian onslaught and ultimately prevent their being overrun, we cannot predict whether something similar may happen to us. Some major destabilizing factor could enter the picture and drastically change the situation. As it looks now, there are basically two types of event that can reshuffle planetary stability: environmental degradation or major armed conflict. These two possibilities are not mutually exclusive.

Environmental disaster can come from different directions. First, there is always the possibility of a comet, asteroid or meteorite hitting the earth. In the 1998 disaster movie *Armageddon*, an asteroid the size of Texas is on a direct collision course with the Earth. Fortunately, a deep-core drilling team is able to split the asteroid to bring it off course before disaster can strike. The possibility of an impact by an extraterrestrial rock is not remote. Indeed, many such objects enter the Earth's atmosphere each day, but virtually all are tiny, just a few milligrams each. Only the largest ones ever reach the surface to become meteorites. Hence, the possibility that such objects will wipe out all life on Earth and destroy the planet is remote. Nevertheless, the impact of a very large comet or asteroid probably caused the extinction of the dinosaurs 65 million years ago. If such an object were to hit the United States, it could easily destroy its economic life and therefore its power. This could destabilize the current world system, leading to renewed military conflicts with different states vying for dominance. At some point this could result in a new 'postindustrial' revolution with some dramatic, new technology to alter society.

While asteroid impacts are highly unlikely events, there are other much less unlikely ways to significantly affect global society. The best known are nuclear holocaust and climate change, both the subject of a lively tradition in literature and film. We all recognize the images of a blighted Earth with civilization crumbling, empty cities with all humans driven back to the land. An accidental nuclear war is of course much more likely than a large meteorite impact as a cause of destroying part of the earth. Large numbers of nuclear weapons are held in military arsenals and some nuclear material is in circulation that could accidentally or through terrorist plots be used against one or more nations. Their deliberate use in an armed conflict will be discussed further below. Either way, the result could easily destroy the balance of global peace. Also in this case it could lead to a society again consisting of vigorously competing states promoting further technology development.

Probably the least unlikely destabilizing event would be global warming. In Marcel Theroux's novel, *Far North*, civilization has been reduced by global warming to pre-industrial levels of technology, with the people pushing north as climate change takes its toll, making sparsely populated areas like the Siberian tundra more livable than lawless cities[182]. There are only remnants of civilization as we know it. Without discussing global warming in intricate detail, as it is beyond the topic of this book, there is currently sufficient evidence to at least make disaster through floods, hurricanes and extreme heat more than a remote possibility. This could indeed make large parts of the earth unlivable, while simultaneously greatly improving the climate in other areas. Like giant meteorites and nuclear accidents, global warming could certainly destabilize world society. In his book *Collapse: How Societies Choose to Fail or Succeed*, Jared Diamond shows how environmental problems caused the demise of some civilizations of the past[183]. The take-home lesson is that in spite of our apparently inexhaustible wealth and sophisticated technology, the same trap could bring us down as well.

While at this stage it is impossible to accurately predict if global warming is a threat that can lead to our collapse, it is certain that human activities are increasingly changing the composition of the atmosphere. Burning fossil fuels has increased atmospheric carbon dioxide concentrations by about 35% since the start of the industrial revolution. Ongoing deforestation is also a key factor since green plants remove carbon dioxide from the atmosphere by photosynthesis. While the increase in carbon dioxide is a fact, its consequences are less clear. What is clear, however, is an increase in the average temperature of the air and oceans of the Earth since the mid-20th century. The global surface temperature increased by almost one degree Celsius. The Intergovernmental Panel on Climate Change (IPCC) concludes that this is mainly due to the increasing greenhouse gas concentrations (mostly carbon dioxide) resulting from burning fossil fuels and deforestation[184]. Although some skeptics still vehemently contest a greenhouse effect, higher temperatures as a result of increased carbon dioxide concentrations is a well-known physical phenomenon, and only on logical grounds it is therefore highly unlikely that the IPCC is wrong in its assessment. An increase in global temperature will cause a continuing retreat of glaciers, permafrost and sea ice and a rise in sea levels. It may change the amount and pattern of precipitation and probably lead to the expansion of subtropical deserts and other ad-

182 Theroux M. *Far North: A Novel.* Farrar, Straus and Giroux, New York, 2009.

183 Diamond J. *Collapse: How Societies Choose to Fail or Succeed.* Penguin, New York, 2005.

184 IPCC Fourth Assessment Report: "Climate Change 2007". Core Writing Team, Pachauri, R.K. and Reisinger, A. (Eds.), IPCC, Geneva, Switzerland.

verse effects, including increases in the intensity of extreme weather events. Therefore, if we assume that the IPCC is right, and again this is the most logical assumption, chances of global warming-driven devastation are not as unlikely as asteroids or nuclear disasters. While collapse cannot be ruled out, society should be able to prevent the worst from happening by taking relatively simple measures for which no major new inventions are needed. As pointed out in Part One, we already have the necessary technology in place to completely substitute alternative energy sources for fossil fuels. Like the Chinese in the 18th century, when they managed to grow new crops and further improve agriculture to feed an ever increasing population, we should be able to implement wind and solar power generation and electric cars when the need is there. Indeed, never before have governments been so capable of forcing their will on society by new regulations and subsidies.

Finally, while global political stability is probably at its highest level ever, is a world-wide military conflict entirely inconceivable? The 2008/2009 severe economic recession could trigger political unrest equal to that seen during the 1930s. However, unlike society in those days we now have a much higher level of government jobs to soften any adverse impact and our financial and economic tools are incomparably more sophisticated. Indeed, the whole argument of this book is that society today is encapsulated by layers of regulation reflecting increasingly good and powerful governance. In the absence of such a safety net we would surely have experienced a repeat of the great depression with all its destabilizing effects. Under the circumstances the effects were benign, which indicates that while governments remain imperfect they have now become reliable pillars of society. So, it should be considered highly unlikely that economic upheaval alone is sufficient to cause a departure of the successful economic globalization model or even lead to world-wide armed conflict. But what about old fashioned power play between new and upcoming super states?

One possible scenario is what the late Samuel Huntington called a 'clash of civilizations'. In his 1996 book *The Clash of Civilizations and the Remaking of World Order*, Huntington predicted competition and conflict according to fault lines determined by religious and cultural traditions[185]. While world-wide conflict on this basis is extremely unlikely, the dominance of the United States and its allies as a geopolitical stabilizer will probably come to an

185 Huntington SP. *The Clash of Civilizations and the Remaking of World Order*. Touchstone, New York, 1996.

end. This is accentuated by the emergence of China, India and Brazil as new major global players, similar to the rise of Germany in the 19th century and the United States itself in the early 20th century. While this will transform the geopolitical landscape, its impact is unlikely to be as dramatic as those of the previous two centuries. Theoretically, the new powers could usher in a new set of international alignments, potentially marking a definitive break with some of the post-World War II institutions and practices. However, an abrupt reversal of the process of globalization would not be to their advantage and in fact prevent their further rise. This is very clear from the restraint shown by China to Taiwan and India to Pakistan. Both rising powers clearly benefit from the status quo and will do nothing that endangers their situation. Yet how China and India will exercise their growing power is of course uncertain.

At the moment, the most likely scenario appears to be the one sketched by Fareed Zakaria in his book *The Post-American World.* Zakaria describes a world in which the United States will slowly lose its dominance of the global economy and will no longer orchestrate geopolitics or overwhelm cultures[186]. He sees this as a positive process in which it is economic growth that produces political confidence and national pride, and not military or political power. This viewpoint is entirely in line with the thesis of this book: that increasingly sophisticated governance has spread like wildfire. Hopefully, it will focus no longer on political dominance, but more and more on the well being of its citizens, even on citizens of the world as a whole. This would be the outcome of a process that began in parts of Asia, the Mediterranean and Europe in the late Middle Ages, was carried by England through the 18th and 19th centuries, brought to new heights in the US, and finally spread from there in our own times.

Can we be successful and still innovative?

The examples of the Roman and Chinese empires tell us that technology decline does not necessary mean the end of empire. Like them, we also are likely to have our best centuries ahead of us. It does mean, however, that we will never again see major breakthroughs that will change our lives. The energy companies, the car and airline industry, the computer industry, the household appliance business, medical industry, all have matured and are not likely to make major investments in something new. Current indus-

186 Zakaria F. *The Post-American World.* Norton, New York, 2008.

try strives for the same objective, that is, to create the lowest-cost business model that will achieve the highest efficiency. That leaves no room for flying cars or nuclear-powered hypersonic planes capable of vertical take-off and landing. How to get that back and still remain a stable, safe society?

There are at least three major obstacles to technological progress: regulatory constraints, advanced business models that effectively seek the highest profit at the lowest possible risk, and a lack of capital. In a sense, these same obstacles were encountered by previous, successful societies in their glory days. However, there is one major difference between us and them, which may allow us at least theoretically to address each of these issues. The bureaucrats of the empires of the past had only limited insight into the constraints that dampened innovativeness in their societies. Most of them would not even have been aware that technological progress in the past had been responsible for their current success. But even had they been aware of that, the sophisticated tools necessary to guide their people back onto a road of innovation were simply not there. This situation may be different today.

First, let's look at regulatory constraints. Sometimes innovation is obstructed by the use of law in response to popular feeling among certain strata of society (by no means limited to action groups), based on perceived safety issues. Many such regulations are well-intended, but the unintended or unobserved consequences of regulation may reduce innovation unnecessarily, through broad disincentives for adopting new, breakthrough inventions. Innovation to pursue such inventions is inherently risky anyway, and it requires the promise of great rewards. Investment decisions depend crucially on the expected return on the investment. If the regulatory framework increases the general risk, then the incentive to innovate becomes even less. By dampening expectations of future returns, regulation saps the incentive to deploy new inventions. Disincentives lead to shortfalls in human capital and research due to underinvestment. Indeed, if investing in existing technology, for example, to increase worker efficiency (even if this limits their flexibility) will pay off better than adopting a new invention, the new technology will not be pursued. Therefore, when new, fragile technology is subject to stringent regulation its further innovation will be short-changed.

Hence, it is important to balance the risks of developing new technology against the risks of not developing it. Where the legal environment diminishes the space available for technological progress, one could imagine a government office using advanced computer models to assess societal effects of regulatory and legal frameworks and to re-balance them so as to allow a maximum of innovation at a minimum of societal risk. This is difficult because much of current regulation is politicized. For example, it is relatively easy to misuse safety regulation for trade-distorting measures among states. But regulation is also misused by incumbent industries to defend old business models threatened by the new. In the past, this happened as well, but to keep a product off the market by promoting fear and misinformation was difficult in a society still inherently unstable, without an established system of government regulations and a public that was not easy to reach and had little political clout. Nowadays, it is incomparably easier to mobilize an angry mob through cable TV and the internet. Hence, the new framework should not only be transparent, science-based, flexible (with new evidence quickly incorporated) and non-discriminatory, but it should also provide legal security for organizations trying to introduce new products and processes to the market. The system should not seek zero risk, as this is unattainable in the real world. Regulatory review should seek to establish that novel products will be as safe as others in the marketplace. In making this evaluation, regulators must take into account both the harms caused by present practices as well as opportunity costs, the potential benefits that would be lost by non-adoption.

The second major hindrance of technological progress is the industrial business model itself, which is based on short-term thinking and reluctance to take risks. This is even true for most venture capitalists and entrepreneurs. Even if regulatory hindrances are reduced, stimulating industry to innovate is difficult. Indeed, if one thinks about it, almost all major inventions from the past were counterintuitive, almost by definition. Had they been subjected to the kind of rational reviews they would nowadays get, they would never have made it. Some explosions of innovativeness took place in highly unstable times, with enormous aggregates of capital (private or centralized) and an almost complete lack of public control. Examples are China's Warring States period, when the kings of several highly competing states sought to maximize their resources at all costs, and 19th century Britain and the US with their powerful industrial magnates trying to do the same thing for their own capitalist empires. Those periods were highly creative, not only in industrial innovation but also in producing works of art. But they also belong to the most violent maelstroms of irrationality the world has ever seen.

How to mimic such 'irrational exuberance' without paying the price of war, social injustice and unsafe environments? For one thing, governments could finance some deliberate irrationality by providing investment funds to pursue innovation in areas private industry will no longer go. Large, government-controlled funds already exist in many countries in the form of sovereign wealth funds, i.e., state-owned investment funds composed of financial assets, such as stocks, bonds, property, and precious metals. Their proceeds could be used for investing in risky projects aiming at the implementation of major breakthrough inventions. It is generally assumed that innovation needs support in its early stages. The assumption is always that once there is 'proof of principle' the rest will follow. But the problem of course is not limited to those earlier stages. Typically, needed investments are orders of magnitude higher later on.

In his book *The Rise and Fall of American Technology*, Lynn Gref calls this Phase Two of technology development. This is the phase in which the discovery made in Phase One (roughly comparable to the invention phase in my Figure 1.1) is turned into a product, such as a steam engine, a transistor, a new drug etc. According to Gref and others, industry has lost much of its ability or willingness to perform Phase Two research that in the past led to the revolutionary devices we are now so familiar with. As mentioned in Chapter 6, the great corporate labs where scientists could pursue risky projects are no more. Nowadays, risks should be reasonable and manageable, which makes all the sense in the world for those not familiar with how breakthrough technology is really developed, i.e., often accidental, based on intuition, with a lot of luck and at a high failure rate by scientists who are often irrational, arrogant and unreasonable.

Hence, to facilitate real technological breakthroughs we should invest in the later stages of the innovation process and again finance the costly and risky Phase Two research. This means that governments should do clinical trials, build plastic cars, develop new rockets and space ships, etc. And they should also provide opportunity for new technology to have a reasonable chance of adoption. Unfortunately, not only did we see the decline of the corporate lab, but Phase Two research is also declining at the Government laboratories, such as the Department of Defense, NASA, the Department of Energy, the Jet Propulsion Laboratory. For example, Gref describes the accomplishments of the latter, from the first guided ballistic missiles, image processing techniques, the clean room (widely used in hospitals and chip manufacturing) and solar arrays to GPS technology. Many of the same people who extensively use all this technology on a daily basis want these labs closed to save taxpayer dollars! It will be difficult to turn the tide and realize that private industry will never take over this role from government, but it is not impossible.

A good example is the National Institutes of Health's recent creation of a center for drug development, the *National Center for Advancing Translational Sciences*. As I explained in Chapter 4, the development of new therapies is typically left to the pharmaceutical industry, as part of the capitalist system. Unfortunately, industry is not developing new drugs fast enough, with most of the 'new' ones merely slight variations of older ones. The establishment of a drug development operation by the government is a sign of mounting frustration that all the great scientific discoveries in the biomedical sciences stubbornly refuse to be translated into new therapies[187]. And this is not because basic scientists have lost their golden touch. Indeed, as long as it is not stem cells or genetically engineered food, basic science is still reasonably free from the regulatory morass that can trap those who try to turn an invention into something useful. Increased concerns about risk and the high costs associated with understanding and addressing these risks to everybody's satisfaction has reduced incentives to develop new therapies. A government-funded center could prove critically important to get things going again. Once the center's activities have greatly reduced the perceived risks, the new therapy can be transferred to private industry, which will then faithfully make further improvements. Hence, the solution to our problem is indeed the government, albeit in partnership with industry. Since governments have the power of the purse, this would also solve the third obstacle to innovation, i.e., lack of capital.

As mentioned in the 2005 report *Rising above the gathering storm* published by the National Academies, federal funding of research in the physical sciences was 45% less in 2004 than in 1976[188]. In a 2009 article in *Business Week*, Michael Mandel notes that of the trillions of dollars flowing into the US over the last decade, most went into government bonds and housing (including

187 Harris G. "Federal research center will help develop medicines." *The New York Times* January 22, 2011.

188 Rising Above the Gathering Storm. www.nationalacademies.org.

subprime mortgages) instead of research and development[189]. All that cheap money was apparently unable to find a destination in the technology sector. While, optimistically, Mandel merely sees this as a temporary lull in innovation, he provides very little concrete evidence for an upswing other than a few examples of biotechnology and IT products, none qualifying as anything close to a major invention.

Hence, a major hindrance nowadays for technological progress is a lack of capital. Not a lack of capital per se, but a lack of willingness to spend large amounts of it on technology development and implementation. In the past, even as recent as the 1960s, a major part of technology development was funded by or on behalf of the military. As explained in Part Two of this book, competitiveness among states has always been a major driver of technological progress and capital was always available for strengthening the state. This included not only the development of new weapons but also programs that had the potential of delivering new weapons, such as the space program. Even if these programs did not result in the anticipated new weapons, they usually had spin-offs that brought us many novel tools and processes. After the end of the cold war, huge defense budgets are no longer available to finance unbridled innovation. Also, conducting colonial adventures to provide a source of extraordinary wealth that could be used for industrialization, as was the case for 18th-century Britain, is no longer an option. New government programs appear to be the only way to get the funds for such major technology programs as futuristic transportation systems, compact cities, robotics, IT infrastructure and space programs.

To ask the question why we need these ambitious programs when we are fine with what we have now, points accurately at the reason why technological progress has stalled in the first place. It is rational thinking, so characteristic for a successful, stable society, that blocks the successful implementation of new macro-inventions. The private sector will certainly not develop ambitious technology programs on its own, which even in the past they only did with massive government support. Hence, the ambitious projects need to come from the government in partnership with private industry. Some will argue that governments are inherently unsuitable to drive technological progress, but one only needs to point towards the Soviet and US (government!) space programs to discard this as utterly untenable.

189 Mandel M. "The failed promise of innovation in the US". *BusinessWeek*, June 3, 2009.

Those programs were breathtaking in their boldness and were executed superbly. It was also the government that developed the first nuclear power reactors. And it was the racing airplanes built and financed by the government in the 1920s that gave rise to a new generation of fast and streamlined airplanes heralding the glorious future of the airline industry[190]. The development of the technology for transmitting data online was made possible by two decades of government support. The Global Positioning System (GPS), a macro-invention now used by millions of motorists, was invented by the U.S. Department of Defense (D.O.D.) at the cost of twelve billion taxpayer dollars. And of course the internet itself, now the world's primary communication network, started as a US Defense Department project. Hence, we need to abandon the idea that everything in society needs to come from private initiative. The government often has a role no company will play on its own.

Many questions remain. For example, is it really possible to inspire innovation and yet not make security and safety a tertiary concern? And is it possible to create demand for new, life-changing technology in a top-down approach with the government specifying the minutiae of policy? Who will decide which new technology program will be adopted? Shouldn't progress come from critically thinking, responsible, clever individuals, rather than from government-organized committees? Providing answers to these questions will be critical for the continuation of a unique period in human history, which for the first time offers us a glimpse of what easily could become the first global civilization. Gaining control over technology development will help us to shape the modern world and our future.

190 Gough M. *Racing Rules*. In preparation.

Table 1: List of macro-inventions since 10,000 BC.

	Approximate time of invention	Nature of invention	Place of the invention
1	10,000 BC	Sawing and harvesting plants	Middle East
2	10,000 BC	Animal husbandry	Middle East
3	9,500 BC	Sickle	Middle East
4	8,000 BC	Quern-stone	Middle East
5	7,000 BC	Brick structures	Middle East
6	6,000 BC	Pottery	Japan
7	6,000 BC	Scratch plow	Middle East
8	6,000 BC	Boat	Middle East
9	5,000 BC	Textiles (wool and linen)	Middle East
10	4,500 BC	Balance scale	Middle East
11	4,500 BC	Metallurgy (copper)	Middle East
12	4,500 BC	Oil lamp	Middle East
13	4,000 BC	Irrigation agriculture	Middle East
14	4,000 BC	Monumental architecture	Middle East
15	4,000 BC	Roads	Middle East
16	3,500 BC	Wheel (cart wheel and potter's wheel)	Middle East
17	3,500 BC	Shadow clock	Middle East
18	3,500 BC	Glass	Middle East
19	3,200 BC	Sailing ship	Middle East
20	3,200 BC	Wheeled cart	Middle East
21	3,100 BC	Bronze	Middle East
22	3,000 BC	Saw	Middle East
23	3,000 BC	Sword	Middle East
24	3,000 BC	Papyrus	Middle East
25	3,000 BC	Water pump (Shaduf)	Middle East

26	3,000 BC	Calendar	Middle East
27	2,800 BC	Cuneiform script	Middle East
28	2,800 BC	Economic institutions (banking, insurance, joint-stock ventures)	Middle East
29	2,000 BC	Spoked wheel (horse-drawn battle chariot)	Caucasus
30	2,000 BC	Mathematics	Middle East
31	1,500 BC	Water clock	Middle East
32	1,500 BC	Soap	Europe
33	1,300 BC	Alphabet	Middle East
34	1,300 BC	Iron smelting	Caucasus
35	700 BC	Locks and keys	Middle East
36	700 BC	Chain pump	Middle East
37	700 BC	Coined money	Middle East
38	500 BC	Lacquer	China
39	500 BC	Rotary quern	Europe
40	500 BC	Blast furnace	China
41	450 BC	Crossbow	China
42	400 BC	Double-armed catapult	Middle East
43	400 BC	Blast furnace	China
44	400 BC	Drawloom	China
45	312 BC	Aqueduct	Europe
46	300 BC	Iron plow with mouldboard	China
47	300 BC	Cast iron	China
48	300 BC	Power transmission (gears, cams, chains)	Europe
49	300 BC	Wooden barrel	Europe
50	300 BC	Rotary lathe	Middle East
51	250 BC	Compound pulley	Europe
52	250 BC	Lever	Europe
53	250 BC	Screw Pump	Europe
54	250 BC	Parchment	Europe
55	250 BC	Decimal numeral system	India
56	250 BC	Waterwheel	Middle East

57	200 BC	Astrolabe	Europe
58	200 BC	Rotary winnowing fan	China
59	200 BC	Stirrup	China
60	200 BC	Multi-tube iron seed drill	China
61	150 BC	Crank handle	China
62	100 BC	Force Pump	Europe
63	100 BC	Wheel roller bearings	Europe
64	100 BC	Swiveling front wheels	Europe
65	100 BC	Dome	Europe
66	100 BC	Horseshoes	Europe
67	100 BC	Concrete	Europe
68	80 BC	Antikythera	Europe
69	80 BC	Hypocaustum	Europe
70	70 BC	Shorthand	Europe
71	50 BC	Deep drilling	China
72	50 BC	Glass blowing	Europe
73	50 BC	Window pane	Europe
74	46 BC	Julian Calendar	Europe
75	30 BC	Hand abacus	Europe
76	9	Caliper	China
77	50	Collapsible umbrella	China
78	50	Wick candle	Europe
79	100	Book (codex)	Europe
80	100	Paper	China
81	100	Shears/scissors	Europe
82	100	Harvesting Machine (Vallus)	Europe
83	100	Junk Ship	China
84	100	Scythe	Europe
85	232	Wheelbarrow	China
86	300	Puddling furnace for making steel	China
87	300	Biological pest control	China
88	400	Horse collar	China
89	500	Algebra	India

90	590	Toilet paper	China
91	600	Block printing	China
92	673	Greek fire	Europe
93	700	Windmill	Middle East
94	700	Lateen sail	Middle East
95	700	Porcelain	China
96	725	Liquid-driven escapement	China
97	750	Hydraulic trip hammer	China
98	800	Sternpost rudder	China
99	800	Spinning wheel	China
100	800	Three-field system	Europe
101	900	Gunpowder	China
102	950	Firelance (proto-gun)	China
103	960	Magnetic compass	China
104	984	Canal lock	China
105	1000	Coke	China
106	1000	Tidal mill	Europe
107	1020	Magnifying glass	Middle East
108	1023	Movable type	China
109	1100	Continuous power-transmitting chain drive	China
110	1100	Watertight ship bulkheads	China
111	1100	Paper Money	China
112	1200	Clothing buttons	Europe
113	1200	Multi-spindle spinning wheel	China
114	1200	Chimney	Europe
115	1250	Logarithmic tables	Europe
116	1285	Eyeglasses	Europe
117	1295	Modern glass	Europe
118	1300	Cotton gin	India
119	1300	Mechanical escapement clocks	Europe
120	1328	Sawmill	Europe
121	1350	Double entry bookkeeping	Middle East

122	1400	Four-wheel sprung carriage	Europe
123	1400	Calculus	India
124	1420	Oilpainting	Europe
125	1450	Spring-powered clocks/watches	Europe
126	1450	Caravel ship	Europe
127	1569	Mercator map projection	Europe
128	1590	Compound microscope	Europe
129	1593	Thermometer	Europe
130	1595	Quadrant	Europe
131	1600	Pencil	Europe
132	1608	Refracting telescope	Europe
133	1621	Slide Rule	Europe
134	1640	Airpump	Europe
135	1643	Barometer	Europe
136	1656	Pendulum clock	Europe
137	1664	Level	Europe
138	1712	Steam engine	Europe
139	1714	Mercury thermometer	Europe
140	1717	Diving bell	Europe
141	1747	Rifle	Europe
142	1757	Sextant	Europe
143	1760	Bifocals	America
144	1761	Marine chronometer	Europe
145	1777	Circular saw	Europe
146	1779	Spinning mule	Europe
147	1783	Steel roller	Europe
148	1783	Hot-air balloon	Europe
149	1785	Chlorine bleaching	Europe
150	1787	Power loom	Europe
151	1787	Soda	Europe
152	1787	Steamboat	America
153	1790	Semaphore telegraph	Europe
154	1792	Gas lighting	Europe

155	1796	Vaccination	Europe
156	1797	Precision lathe	Europe
157	1799	Battery	Europe
158	1805	Refrigeration machine	America
159	1810	Tin can	Europe
160	1811	McAdam roads	Europe
161	1814	Steam locomotive	Europe
162	1821	Electric motor	Europe
163	1830	Sewing machine	Europe
164	1831	Dynamo	Europe
165	1832	Mechanical reaper	America
166	1833	Rotary press	America
167	1836	Screw propeller	America, Europe
168	1837	Mechanical shovel (steam)	America
169	1837	Telegraph	America
170	1839	Daguerreotype	Europe
171	1839	Vulcanization of rubber	America
172	1840	Chemical fertilizer	Europe
173	1843	Fax machine	Europe
174	1846	Anesthesia	America
175	1847	Antiseptics	Europe
176	1849	Safety pin	America
177	1852	Gyroscope	Europe
178	1853	Elevator brake	America
179	1855	Bessemer steel process	Europe
180	1855	Aluminum production by electrolysis	Europe, America
181	1855	Safety matches	Europe
182	1856	First artificial dye (aniline purple)	Europe
183	1858	Internal combustion engine	Europe
184	1859	Spectroscope	Europe
185	1859	First oil drill	America
186	1862	Pasteurization	Europe

187	1867	paper clip	Europe
188	1866	Dynamite	Europe
189	1868	Typewriter	America
190	1869	Plastic (celluloid)	America
191	1869	Airbrake	America
192	1869	Bicycle	Europe
193	1874	Washing machine	America
194	1874	Barbed wire	America
195	1875	Electric lightbulb	America, Europe
196	1875	Electric dental drill	America
197	1876	Arc lamp	Europe
198	1876	Telephone	America
199	1876	Microphone	Europe
200	1877	Staples	America
201	1877	Phonograph	America
202	1880	Seismograph	Europe
203	1882	Electric grid	America
204	1883	A-C Induction motor	America
205	1884	Photographic film	America
206	1884	Steam turbine	Europe
207	1885	Mechanical typesetting	Europe
208	1885	Maxim machine-gun	America
209	1886	Dishwasher	America
210	1888	Pneumatic tire	Europe
211	1889	Automobile	Europe
212	1889	Skyscraper	America
213	1891	Escalator	America
214	1892	Farm tractor	America
215	1895	Motion picture	Europe
216	1895	X-ray photographs	Europe
217	1898	Radio	America
218	1899	Aspirin	Europe
219	1900	Zeppelin	Europe

220	1901	Assembly line	America
221	1901	Fluorescent lamp	America
222	1901	Disc brake	America
223	1902	Sheet glass drawing machine	America
224	1902	Air conditioner	America
225	1903	Airplane	America
226	1903	Porcelain dental crown	America
227	1905	Novocain	Europe
228	1907	Precision cast dental fillings	America
229	1907	Electric vacuum cleaner	America
230	1907	Vacuum tube	America
231	1909	Toaster	America
232	1909	Tuned mass damper	America
233	1913	Zipper	America
234	1916	Battle tank	Europe
235	1916	Drone (Remotely Piloted Vehicle)	Europe
236	1926	Rocket	America
237	1927	Television	America
238	1928	Penicillin	Europe
239	1930	Fiber optics	Europe
240	1930	Scotch tape	America
241	1930	Jet engine	Europe
242	1931	Electron microscope	Europe
243	1934	Tape recorder	Europe
244	1935	Radar	Europe
245	1936	Helicopter	Europe
246	1937	Photocopier	America
247	1938	Ballpoint pen	Europe
248	1938	Compacting garbage truck	America
249	1938	Nylon toothbrush	America
250	1939	Electronic digital computer	America
251	1939	DDT	Europe
252	1940	Color television	America

253	1942	Nuclear reactor	America
254	1942	Night vision technology	Europe
255	1944	Kidney dialysis machine	Europe
256	1945	Atomic bomb	America
257	1946	Microwave oven	America
258	1946	Assault rifle (Kalashnikov)	Europe
259	1947	Transistor	America
260	1948	Holography	Europe
261	1949	Bar Codes	America
262	1949	Air turbine dental drill	Australia
263	1950	Optical character recognition	America
264	1952	Carbon nanotubes	Russia
265	1952	Electronic hearing aid	America
266	1954	Oral contraceptives	America
267	1954	Solar cell	America
268	1956	Videotape recorder	America
269	1956	Magnetic disk drive	America
270	1957	Cochlear implant (bionic ear)	America
271	1957	High-level computer language (FORTRAN)	America
272	1957	Satellite	Europe
273	1958	Laser	America
274	1958	Anti-hypertensive drugs	America
275	1958	Artificial network (Perceptron)	America
276	1959	Hydrogen fuel cell	Europe
277	1959	Integrated circuit	America
278	1960	Tokamak	Russia
279	1961	Human space travel	Europe
280	1961	Computer time-sharing	America
281	1962	Implantable pacemaker	America
282	1964	Electret microphone	America
283	1964	Computer mouse and graphical user interface	America

284	1964	Automated facial recognition	America
285	1964	Geostationary satellite	America
286	1965	Compact disc	America
287	1965	Titanium dental implants	Europe
288	1966	Kevlar	America
289	1969	Artificial hip and knee joint systems	Europe, America
290	1969	Charge-coupled device (CCD)	America
291	1970	Robot	America
292	1971	Email	America
293	1971	Excimer laser	Europe
294	1971	Microprocessor	America
295	1971	LCD display	America, Europe
296	1972	Internet	America
297	1972	Microcomputer	America
298	1972	Precision-guided smart bomb	America
299	1972	Computerized Axial Tomography (CAT) scan	Europe, America
300	1972	Fluorescence activated cell sorting	America
301	1973	Gene cloning	America
302	1973	Cellular phone	America
303	1976	Cray supercomputer	America
304	1976	Statins	America
305	1977	DNA sequencing	Europe
306	1977	Lithium ion battery	America
307	1977	Magnetic resonance imaging	America
308	1978	Global positioning system	America
309	1980	Automatic implantable defibrillator	America
310	1981	Scanning tunneling microscope	Europe
311	1982	Artificial heart	America
312	1983	PCR	America
313	1985	File sharing (FTP)	America

314	1985	Robotic surgery	Europe
315	1985	DNA fingerprinting	Europe
316	1985	First practical speech recognizer	America
317	1986	Giant magnetoresistance	Europe
318	1986	High-temperature ceramic superconductor	Europe
319	1988	Genetically engineered animal	America
320	1989	Microarray	America
321	1989	World Wide Web	Europe
322	1993	High-intensity light-emitting diode (LED) lamp	Japan, America, Europe
323	1994	Coronary artery stent	America
324	1994	First recombinant DNA-derived whole food product (the FlavrSavr tomato)	America
325	1994	First search engine (WebCrawler)	America
326	1995	Positron Emission Tomography (PET) scan	America
327	1995	Miniature video cameras	America
328	1995	Plasma window	America
329	1996	First cloned mammal	Europe
330	1996	Instant messaging	America
331	1997	Gas-electric hybrid car	Asia
332	1998	Erectile dysfunction drug (viagra)	Europe
333	2000	Horizontal fracking	America
334	2003	Scramjet	America
335	2003	Massively parallel sequencing	America
336	2004	E-book reader (Sony)	Asia
337	2006	Inducible pluripotent stem cells	Asia

Note: I used multiple sources for generating the list of macro-inventions, including Joel Mokyr's *The Lever of Riches*, Bunch and Hellemans' *The History of Science and Technology*, McClellan and Dorn's *Science and Technology in World History*, Wikipedia's timeline of historic inventions, and a great many newspaper and magazine articles, web sites, and some other books as well. I realize that the list is incomplete and subjective.

Acknowledgments

This book is the result of a personal life-long interest in science and technology. It is very different from the scientific articles and the occasional books I usually write. In my professional work most of the writing is for insiders and can be done following a certain formula with a lot of standard phraseology. With this book I tried to formulate my thoughts in a way that is accessible to average educated readers, whether they are scientists or not. As a non-specialist in most of the many topics covered in this book, I have been unable to use the rigorous tools of the scientific method to prove the points I am trying to make. But then, how much of this is provable? One of the pleasures in writing this book was the opportunity to explore a wide range of topics, trying to explain and interpret some of the world's facts on technological progress now and in history. I readily admit that much of my reasoning is necessarily based on inconclusive evidence, but I hope that this nevertheless could serve as a framework for further study of the subject.

There are a lot of people to thank. First of all, I lay no intellectual claim on any of the observations I describe. Therefore, I thank all the authors of books, magazine and newspaper articles, blogs and web sites, whose work I cited and some I surely forgot to cite, for the inspiration, many facts and the ideas they provided me. They provided me with the facts and background information that are indispensable for writing a book like this. I thank Jeff Lewis for Fig. 6.1 and Elsevier for permission to reproduce Fig. 1.2.

I am grateful to friends and colleagues who endlessly discussed with me many of the issues raised in the book. Most of all I thank my friend and colleague Yousin Suh, who is really the one who stimulated me to write this book. I vividly remember our passionate discussion while driving along a German autobahn on whether or not there was a need to move towards greener technology for raising energy. Not a good place to be passionate about something!

My friend Michael Gough, himself a writer, was kind enough to read a large part of the manuscript in detail and give me tremendous input and subject for thought. I am extremely grateful for that. I would also like to thank my colleagues Ganjam Kalpana and Jack Lenz for sharing the interesting experiences with me that are recounted in Chapter 4. I also thank Brent Calder, who prepared all the figures and the table and also gave me valuable advice and new ideas on many aspects of the book. I thank my wife, Claudia Gravekamp, for the many discussions we had over dinner on the various topics dealt with in the book. Finally, I am extremely grateful to the Algora Publishing team for patiently working with me in turning my manuscript into a real book, a process that I greatly underestimated.

Most of all, I thank my father, Leen Vijg, who was the one to awaken my interest in the world around us, in history and in science and technology.

Bibliography

Part One – Broken Promises

Akst, J. "EU Trial Rules Stall Research." *The Scientist* (2009).

Alarcon-Vargas, D., Z. Zhang, B. Agarwal, K. Challagulla, S. Mani, and G. V. Kalpana. "Targeting Cyclin D1, a Downstream Effector of INI1/hSNF5, in Rhabdoid Tumors." *Oncogene* 25, no. 5 (Feb 2, 2006): 722-734. doi:10.1038/sj.onc.1209112.

Arntzen, CJ. "GM Crops: Science, Politics and Communication." *Nature Reviews Genetics* 4 (2003): 839-843.

Ashdown SP, Loker S, Schoenfelder K, Lyman-Clarke L. "Using 3D Scans for Fit Analysis." *J. of Textile and Apparel, Technology and Management* 4 (2004).

Asimov, I. *The Foundation Trilogy*. Doubleday, 1951.

———. *The Gods Themselves*. Bantam, 1972.

———. *I, Robot*. Doubleday, 1950.

Atzmon, G. "Lipoprotein Genotype and Conserved Pathway for Exceptional Longevity in Humans." *PLoS Biol* (2006): 113.

Balter, M. "Was North Africa the Launch Pad for Modern Human Migrations?" *Science* 331, (2011): 20-23.

Brumfiel, G. "Just Around the Corner." *Nature* no. 436 (2005): 318-320.

Bunch, B., Hellemans, A. *The History of Science and Technology*. New York: Houghton Mifflin, 2004.

Clarke, AC. *2001: A Space Odyssey*. New York: Penguin, 1986.

Denning, PJ (editor). *The Invisible Future*. New York: McGraw Hill, 2002.

Diamond, J. *Collapse: How Societies Choose to Fail Or Succeed*. New York: Viking, 2005.

Gallagher RM. "Balancing Risks and Benefits in Pain Medicine: Whither Vioxx." *Pain Medicine* no. 5 (2004): 329-330.

Gelernter, D. *1939: The Lost World of the Fair*. New York: Free Press, 1995.

Goklany, I. *The Improving State of the World: Why We're Living Longer, Healthier, More Comfortable Lives on a Cleaner Planet.* Washington, DC: Cato Institute, 2007.

Gref, LG. *The Rise and Fall of American Technology*. New York: Algora Publishing, 2010.

Haber, DA. "The Evolving War on Cancer." *Cell* no. 145 (2011): 19-24.

Henderson, LW. "Obituary: A Tribute to Willem Johan Kolff, M.D. 1912 - 2009." *J Am Soc Nephrol* 20 (2009): 923-924.

Howell, K. "Exxon Mobil Bets $600 Million on Algae." *Scientific American* (July 14, 2009).

Huebner, J. "A Possible Declining Trend for Worldwide Innovation, Technological Forecasting & Social Change." (2005): 980-986.

Izbicka, E. "Distinct Mechanistic Activity Profile of Pralatrexate in Comparison to Other Antifolates in in Vitro and in Vivo Models of Human Cancers." *Cancer Chemother Pharmacol* (2009): 64-993-999.

Jaffe, G. "Regulatory Slowdown on GM Crop Decisions." *Nat Biotechnol* (2006): 24-748-749.

Kaiser, J., Vogel, G. "Controversial Ruling Throws U.S. Research into a Tailspin." *Science* no. 329 (2010): 1132-1133.

Kaku, M. *Physics of the Impossible.* New York: Doubleday, 2008.

Kean, S. "Making Smarter, Savvier Robots." *Science* no. 329 (2010): 508-509.

Krieger, M. "The "Best" of Cholesterols, the "Worst" of Cholesterols: A Tale of Two Receptors." *Proc. Natl. Acad. Sci USA* no. 95 (1998): 4077-4080.

Kurzweil, R. *The Singularity is Near: When Humans Transcend Biology*. New York: Viking, 2005.

Leaf, C. "Why We're Losing the War on Cancer [And How to Win It]." *Fortune* no. 149 (2004): 76-97.

Licklider, JCR, Taylor, RW. "The Computer as a Communication Device." *Science and Technology* no. 76 (1968): 21-31.

Lu, X., McElroy, MB., Kiviluoma, J. "Global Potential for Wind-Generated Electricity." *Proc. Natl. Acad. Sci USA* no. 106 (2009): 10933-10938.

Malthus, TR. *An Essay on the Principle of Population.* (1798) Winch D. (Ed). Cambridge: Cambridge University Press, 1992.

Mann, A. "NASA Human Space-Flight Programme Lost in Translation." *Nature* no. 472 (2011): 16-17.

Mann, T. *The Magic Mountain.* New York: Vintage, 1996.

Meadows, DH. Meadows, DL, Randers, J, Behrens III, WW. *The Limits to Growth.* New York: Universe Books, 1972.

Miklos GLG. "The Human Cancer Genome Project- One More Misstep in the War on Cancer." *Nature Biotechnology* no. 23 (2005): 535-537.

Mittra, I. "Why is Modern Medicine Stuck in a Rut?" *Perspect Biol Med* no. 52 (2009): 500-517.

Modeling Human Risk: Cell & Molecular Biology in Context. NASA and LBNL Publication #40278. Springfield, VA (1997).

Mokyr, J. *The Lever of Riches: Technological Creativity and Economic Progress.* New York: Oxford, 1990.

———. "Punctuated Equilibria and Technological Progess." *American Economics Review* no. 80 (1990): 350-354.

Moore, GE. "Cramming More Components Onto Integrated Circuits." *Electronics* no. 38 (1965).

Morice, AH. "The Death of Academic Clinical Trials." *The Lancet* no. 361 (2003).

Naldini, L. "Inserting Optimism Into Gene Therapy." *Nature Medicine* 12, (2006).

Negroponte, N. *Being Digital.* New York: Vintage, 1996.

Olsen, M. *The Rise and Decline of Nations.* New Haven: Yale University Press, 1982.

Petkantchin, V. *Risks and Regulatory Obstacles for Innovating Companies in Europe.* Bruxelles: Institut Economique Molinari, 2008.

Roeb, M., Muller-Steinhagen, H. "Concentrating on Solar Electricity and Fuels." *Science* no. 329 (2010): 773-774.

Sagan, C. *Pale Blue Dot: A Vision of the Human Future in Space.* New York: Ballantine, 1994.

Schiermeier, Q. "German Universities Bow to Public Pressure Over GM Crops." *Nature* no. 453 (2008): 263.

Schmid, EF. Smith, DA. "Is Declining Innovation in the Pharmaceutical Industry a Myth?" *Drug Discovery Today* 10, (2005): 1031-1039.

Schwartz, RS. "The Politics and Promise of Stem-Cell Research." *The New England Journal of Medicine* 355, (2006): 1189-1191.

Seife, C. *Sun in a Bottle*. New York: Viking, 2008.

Silverberg, R. *Shadrach in the Furnace*. University of Nebraska Press, 1976.

Smart, J. "Measuring Innovation in an Accelerating World: Review of "A Possible Declining Trend for Worldwide Innovation"." *Jonathan Huebner, Technological Forecasting & Social Change* 72, (2005): 988-995.

Snijders, C., Tazelaar, F., Batenburg R.,. "Electronic Decision Support for Procurement Management." *Journal of Purchasing and Supply Management*. Vol. 9, 191-198, 2003.

Stewart, DJ. "Equipoise Lost: Ethics, Cost, and the Regulation of Cancer Clinical Research." *J. Clinical Oncology* 28 (2010): 2925-2935.

Thaxton, CS, Daniel, WL. Giljohann, DA, Thomas AD, Mirkin CA. "Templated Spherical High Density Lipoprotein Nanoparticles." *J Am Chem Soc* 131, (2009): 1384-1385.

Turing, AM. "Computing Machinery and Intelligence." *Mind* 59, (1950): 433-460.

Vane, J. "The History of Inhibitors of Angiotensin Converting Enzyme." *J Physiol Pharmacol* 50, (1999): 489-498.

Verne, J. *From the Earth to the Moon* (1865). New York: Barnes & Noble Books, 2005.

———. *Paris in the Twentieth Century: Jules Verne, the Lost Novel*. New York: Ballantine, 1996.

Vijg, J. Campisi, J. "Puzzles, Promises and a Cure for Ageing." *Nature* 454, (2008): 1065-1071.

Von Hippel, E. *Democratizing Innovation*. Cambridge: MIT Press, 2005.

Williams, DA., Baum, C. "Gene Therapy- New Challenges Ahead." *Science* 302, (: 400-401.

Young, RC. "Cancer Clinical Trials–A Chronic but Curable Crisis." 363, (2010): 306-209.

Part Two – Parallel Worlds

Pompeii: A Novel. New York: Ballantine, 2003.

Bar-Yosef Mayer DE. "Shells and Ochre in Middle Paleolithic Qafzeh Cave, Israel: Indications for Modern Behavior." *J Hum Evol* 56, (2009): 307-314.

Boorstin, DJ. *The Discoverers: A history of man's search to know his world*. New York: Vintage, 1985.

Boorstin, DJ. *The Creators: A history of heroes of the imagination*. New York:

Vintage, 1993.
Braudel, F. *A History of Civilizations.* New York: Penguin, 1995.
———. "Memory and the Mediterranean." (2001).
Bridges, JH. *The 'Opus Majus' of Roger Bacon, Volume l* Adamant, 2005.
Carbeuri, RL. "A Theory of the Origin of the State." *Science* 169 (1970): 733-738.
Clark, G. *A Farewell To Alms: A Brief Economic History of the World.* Princeton: Princeton University Press, 2007.
Cotterell, A. *The Imperial Capitals of China.* London: Pimlico, 2007.
Cummings, B. *Korea's Place in the Sun.* New York: Norton, 2005.
Darwin, C. *The Descent of Man.* London: John Murray, 1871.
De Loo, Ivo & Soete, Luc. "*The Impact of Technology on Economic Growth: Some New Ideas and Empirical Considerations*". Vol. 017 MERIT, Maastricht Economic Research Institute of Innovation and Technology, 1999.
Diamond, J. *Guns, Germs, and Steel.* New York: Norton, 1997.
———. *How Societies Choose to Fail Or Succeed.* New York: Penguin, 2005.
———. "The Worst Mistake in the History of the Human Race." *Discover Magazine* no. May (1987): 64.
Edwards, R. (Translator). Multatuli: *Max Havelaar: Or the Coffee Auctions of the Dutch Trading Company.* London: Penguin, 1960.
Everitt, A. *Augustus: The Life of Rome's First Emperor.* New York: Random House, 2006.
Ferguson, N. *Empire: The Rise and Demise of the British World Order and the Lessons for Global Power.* New York: Penguin, 2002.
Fernandez-Armesto, F. *The Americas* Modern Library, 2003.
Finlayson, C. "Biogeography and Evolution of the Genus Homo." *TRENDS in Ecology and Evolution* 20, (2005): 457-463.
Friedman, TL. *The World is Flat: A Brief History of the Twenty-First Century.* New York: Farrar, Straus & Giroux, 2005.
Gibbon, E. *The Decline and Fall of the Roman Empire.* Vol. 1-3. New York: Knopf, 1993.
Gies, F., Gies, J. *Cathedral, Forge, and Waterwheel: Technology and invention in the Middle Ages.* New York: HarperCollins, 1994.
Gimpel, J. *The Medieval Machine: The Industrial Revolution of the Middle Ages.* New York: Barnes & Noble Books, 1976.
Gomez-Salvador, A. Musso, A., Stocker, M., Turunen, J. "European Central Bank, Occasional Paper Series.", http:/www.ecb.int.
Goodall, J. *The Chimpanzees of Gombe: Patterns of Behavior.* Cambridge: Belknap

Press of Harvard University Press, 1986.

Grant, M. *The Climax of Rome*. London: Weidenfeld, 1968.

Graves, R. (Translator). *The Golden Ass*. New York: Noonday, 1951.

Haddawy, H. (Translator). "Sinbad: And Other Stories From the Arabian Nights." (.

Heather, PJ. *The Fall of the Roman Empire*. New York: Oxford, 2005.

Heer, F. *The Medieval World: Europe 1100 - 1350*. London: Weidenfeld, 1961.

Henshilwood, CS. "Middle Stone Age Shell Beads from South Africa." *Science* 384, (2004): 404.

Henshilwood, CS. Marean, CW. "The Origin of Modern Human Behaviour: A Review and Critique of Models and Test Implications." *Current Anthropology* 44, (2003): 627-651.

Hibbs, DA Jr., Olsson, O. "Geography, Biogeography and Why some Countries are Rich and Others Poor." *Proc. Natl. Acad. Sci USA* 101, (2004): 3715-3720.

Hodges, H. *Technology in the Ancient World*. New York: Barnes & Noble Books, 1970.

Hunt, GR. "Animal Behaviour: Laterality in Tool Manufacture by Crows." *Nature* 414, (2001): 707.

Huntington, SP. *The Clash of Civilizations and the Remaking of World Order*. New York: Touchstone, 1996.

Israel, J. *The Dutch Republic: Its Rise, Greatness and Fall 1477-1806*. New York: Oxford, 1995.

Jacobsen, T., Adams, RM. "Salt and Silt in Ancient Mesopotamian Agriculture." 128, (1958): 1251-1258.

Jardine, L. *Going Dutch: How England Plundered Holland's Glory*. New York: Harper Collins, 2008.

Landes, DS. *The Wealth and Poverty of Nations: Why some are so Rich and some so Poor*. New York: Norton, 1999.

Landes, DS. *Revolution in Time*. Cambridge: Harvard University Press, 1983.

Latham, R. (Translator). *Marco Polo: The Travels*. New York: Penguin, 1958.

Laven, M. *Mission to China: Matteo Ricci and the Jesuit Encounter with the East*. London: Faber and Faber, 2011.

Lev-Yadun, S. "Archaeology: The Cradle of Agriculture." 288, (2000): 1602-1603.

Lin, JY. "The Needham Puzzle: Why the Industrial Revolution did not Originate in China." *Economic Development & Cultural Change* 43, (1995):

269-292.

McNeill, WH. *The Rise of the West: A History of the Human Community* Chicago: University of Chicago Press, 1991 (originally published in 1963).

Mokyr, J. "Chapter 17: Long-Term Economic Growth and the History of Technology." In *Handbook of Economic Growth*, edited by Aghion, P., Durlauf, SN. Vol. 1B: Elsevier, 2005.

———. *The Lever of Riches: Technological Creativity and Economic Progress*. New York: Oxford, 1990.

Mote, FW. *Imperial China 900 - 1800*. Cambridge: Harvard University Press, 1999.

Pachauri, RK., Reisinger, A. "IPCC Fourth Assessment Report: Climate Change." (2007) (http://www.ipcc.ch/pdf/assessment-report/ar4/syr/ar4_syr.pdf).

Pearsall, DM. "From Foraging to Planting." *Science* 313, (2006): 173-174.

Pickstone, J. "Islamic Inventiveness." *Science* 313, (2006): 47.

Radice, M. (Translator). *The Letters of the Young Pliny*. London: Penguin, 1963.

Ronan, CA. *The Shorter Science & Civilisation in China: 1. An abridgement of Joseph Needham's original text.* 1978, Cambridge University Press, Cambridge.

Schama, S. *The Embarrassment of Riches: An Interpretation of Dutch Culture in the Golden Age* Vintage, 1987.

Scott-Kilvert, I. (Translator). Cassius Dio: *The Roman History: The Reign of Augustus.* London: Penguin, 1987.

Solow, RM. "A Contribution to the Theory of Economic Growth." *Quarterly Journal of Economics* 70, no. 1 (1956): 65-94.

Tebbich, S. "Do Woodpecker Finches Acquire Tool-use by Social Learning?" *Proc. R. Soc. Lond. B.* 268, (2001): 2189-2193.

Templeton, AR. "Out of Africa again and again." *Nature* 416, (2002): 45-51.

Theroux, M. *Far North: A Novel*. New York: Farrar, Straus, Giroux, 2009.

Thomas, H. *Rivers of Gold: The Rise of the Spanish Empire, from Columbus to Magellan*. New York: Random House, 2003.

Thorpe, L. (translator), Gregory of Tours. *The History of the Franks*. New York: Penguin, 1974.

Tin-bor Hui, H. *War and State Formation in Ancient China and Early Modern Europe*. New York: Cambridge, 2005.

Van Lawick Goodall, J., Van Lawick Goodall, H. "Use of Tools by Egyptian Vultures, Neophron Percnopterus." *Nature* 212, (1966): 1468-1469.

Weiner, S. "Evidence for the use of Fire at Zhoukoudian, China." *Science* 281, (1998): 251-253.

Westergaard, GC. "The Stone-Tool Technology of Capuchin Monkeys: Possible Implications for the Evolution of Symbolic Communication in Hominids." *World Archaeology* 27, (1995): 1-9.

Whitton, D. (editor: Holmes, D.). The Society of Northern Europe in the Middle Ages, in: *The Oxford History of Medieval Europe.* Oxford, 1988.

Zakaria, F. The Post-American World. New York: Norton, 2008.

Index

K

L

M

N

O

P

Q

R

S